Geographical Aspects of Agricultural Changes in the Ryukyu Islands

Shannon McCune

A University of Florida Book

The University Presses of Florida
Gainesville · 1975

Library of Congress Cataloging in Publication Data

McCune, Shannon Boyd-Bailey, 1913–
Geographical aspects of agricultural changes in the Ryukyu Islands.

(University of Florida social sciences monograph; no. 54)
"A University of Florida book."
1. Agricultural geography—Japan—Okinawa (Prefecture) 2. Agriculture—Japan—Okinawa (Prefecture)—History. I. Title. II. Series: Florida. University, Gainesville. University of Florida monographs: Social sciences; no. 54.
S471.J320455 630'.952'81 75–11729
ISBN 0–8130–0487–X

PRINTED BY THE STORTER PRINTING COMPANY, INCORPORATED
GAINESVILLE, FLORIDA

Acknowledgments

I am very indebted to a large number of persons and organizations who have assisted me in this study. A particular debt of gratitude is owed to many Okinawan farmers, men and women, who willingly discussed their situations and problems with me both while I was serving as Civil Administrator of the Ryukyu Islands from 1962 to 1964 and while I was conducting my field research in the Ryukyu Islands in 1967 and again in 1970. I appreciate very much the assistance given to me in 1970 by two young Okinawan geographers, both trained in the United States, now associated with the University of the Ryukyus: Michihiro Miyagi and Shinzo Shimabukuro. They translated for me in the field and helped me in many other ways. Dr. Miyagi very kindly obtained for me in 1972 preliminary reports from the 1971 Agricultural Census of the Ryukyu Islands and Mr. Shimabukuro sent me copies of the completed reports in 1974.

During the summer of 1970 I held discussions with many officials of the Government of the Ryukyu Islands and of the United States Civil Administration of the Ryukyu Islands. Conversations with geographers, agricultural economists, and others at the University of the Ryukyus were also very helpful to me. Local government officials, officers of branches of the agricultural cooperative organizations, and managers of sugar mills and pineapple canneries were kind in answering my specific queries for information.

My wife, Edith Blair McCune, and my son, George Blair McCune, have helped me greatly in editing and typing manuscripts for this study. Graduate students in the Department of Geography of the University of Florida, Matthew Igbozuruke, Steven T. Gladin, Glenn M. Fish, and Sioux Remer, have assisted me with the maps and charts. Lennie S. Shaw very kindly typed the final draft.

The Ryukyu Islands Project of which this monograph is an important part would not have been possible without the financial support of the National Science Foundation and the University of Florida. I am deeply thankful for this support. The interest and advice I have received from my colleagues in the Department of Geography of the University of Florida has been appreciated. Thanks must go specifically to the Graduate School of the University of Florida for making possible the publication of this monograph.

Though I have received assistance from many people and organizations, I am solely responsible for the analysis and comments on the data presented in this monograph. My hope is that through this publication more people will be aware of and will be able to improve the situation of the farmers of the Ryukyu Islands as they face a world of great changes.

Contents

GEOGRAPHICAL ASPECTS of AGRICULTURAL CHANGES In the RYUKYU ISLANDS

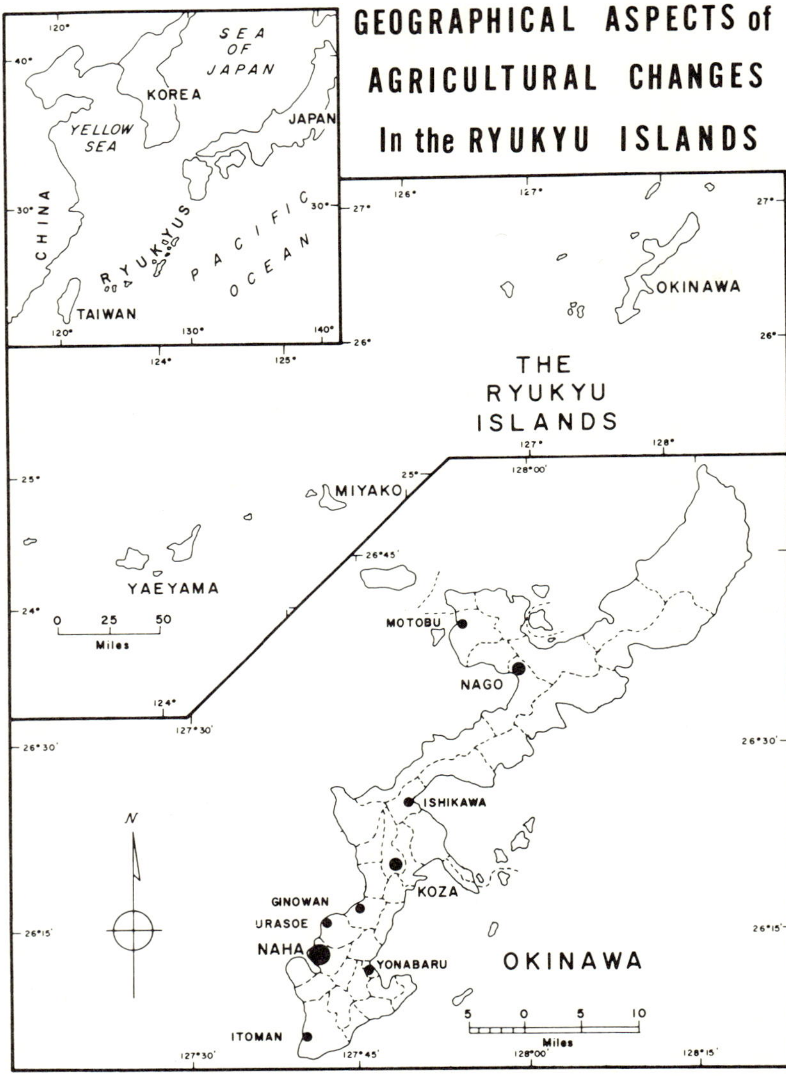

1. Overview

ON May 15, 1972, a significant political change took place in the Ryukyu Islands: the United States Civil Administration and the Government of the Ryukyu Islands gave place to the Japanese prefectural government of Okinawa.[1] This change in sovereignty has had a great impact on the agriculture of the Ryukyu Islands.[2] Agricultural changes in the Ryukyu Islands are not however new phenomena. Through the centuries the introduction of new crops such as sweet potatoes and sugarcane has had a profound impact on the character of agriculture and on the capacity of the islands to support increased population. In recent decades the actions of the United States Civil Administration, the Government of the

1. The name of the Ryukyu Islands has been romanized in many ways. The romanization of the Chinese pronunciation is Liu chiu, but it has been romanized in other ways; two which were widely used are Lew Chew and Loo Choo. The islands as a whole have been called in Japanese the Nansei Retto, literally translated as the Southwest Archipelago, and are referred to as such in many of the official treaties and documents. When the Kingdom of Liu chiu (or Ryukyu in Japanese) was absorbed into Japan in 1879, the islands became administratively a prefecture of Japan and were named for the largest and most significant island, Okinawa Prefecture. After World War II, the area under United States administration included the Amami-Oshima islands to the north of Okinawa; these had been a part of Kagoshima Prefecture prior to that time. In 1953 the Amami-Oshima were given back to Japan and for the area under the United States Civil Administration the term Ryukyu Islands was continued in use. With the reversion to Japan in 1972 the name was changed to Okinawa Prefecture. In this monograph the name generally used is that which was officially in use at the period of time under discussion. It needs to be noted that both as a Japanese prefecture and as an area under United States administration, the islands were divided into three parts for statistical and some governmental purposes: Okinawa and its nearby islands, Miyako and its nearby islands, and Yaeyama, the southwestern islands, including Ishigaki and Iriomote.

2. Much has been written about the reversion of the Ryukyu Islands to Japan in 1972. A brief article on "The Ryukyu Islands: Geographic Aspects of a Change in Sovereignty" was published in *The Journal of Geography* 71 (September 1972) by this author.

Ryukyu Islands, and the Government of Japan have greatly influenced the agricultural situation.

This monograph summarizes the geographical aspects of the major changes in agriculture which have taken place in recent times in the Ryukyu Islands. By studying these changes it may be possible to predict some of the future changes that may take place in Okinawa. With the assistance of grants from the National Science Foundation and from the University of Florida, research on these interesting geographical aspects of agricultural changes in the Ryukyu Islands has been undertaken.[3] Some of the results of this research are presented in this statistical and map study.

Traditionally, the people of the Ryukyu Islands have been farmers and fishermen, operating on a simple, self-sufficient basis. However, during the last two decades service and other industries have grown phenomenally, with a resultant decline in the relative economic importance of agriculture. One of the causes for this change was that after 1950 the Ryukyu Islands became the site of major American military supply and staging bases for American efforts in the Korean War and, later, in the Vietnam conflict. Called the "Keystone of the Pacific," the island of Okinawa was developed as a strategic salient in the American defense against threat and tension in the Far East. Despite the reversion to Japan in May 1972, the strategic military position of Okinawa continues to be of great significance.

There is considerable geographical diversity within the Ryukyu Islands. Central and southern Okinawa have many urban centers as well as extensive American military installations. According to the population census of October 1970, 45.5 per cent of the 945,111 inhabitants of the Ryukyu Islands were concentrated in this small area. The city of Naha alone had a population of 276,380. Naha and the rest of southern Okinawa have been completely rebuilt following the devastation of the Battle of Okinawa, the last great battle of World War II. The urban area continues north from Naha to Koza, a former rural crossroads near which a major Japanese military air base, Kadena Air Base, was built in 1944 and then ex-

3. Twenty-three *Research and Information Papers* of the Ryukyu Islands Project have been prepared in xeroxed form and give an account of some of this research. They also include reprints of scarce and out-of-print material of value in Ryukyuan studies. Copies of these papers are available in various research libraries. A list of these is given in Appendix A.

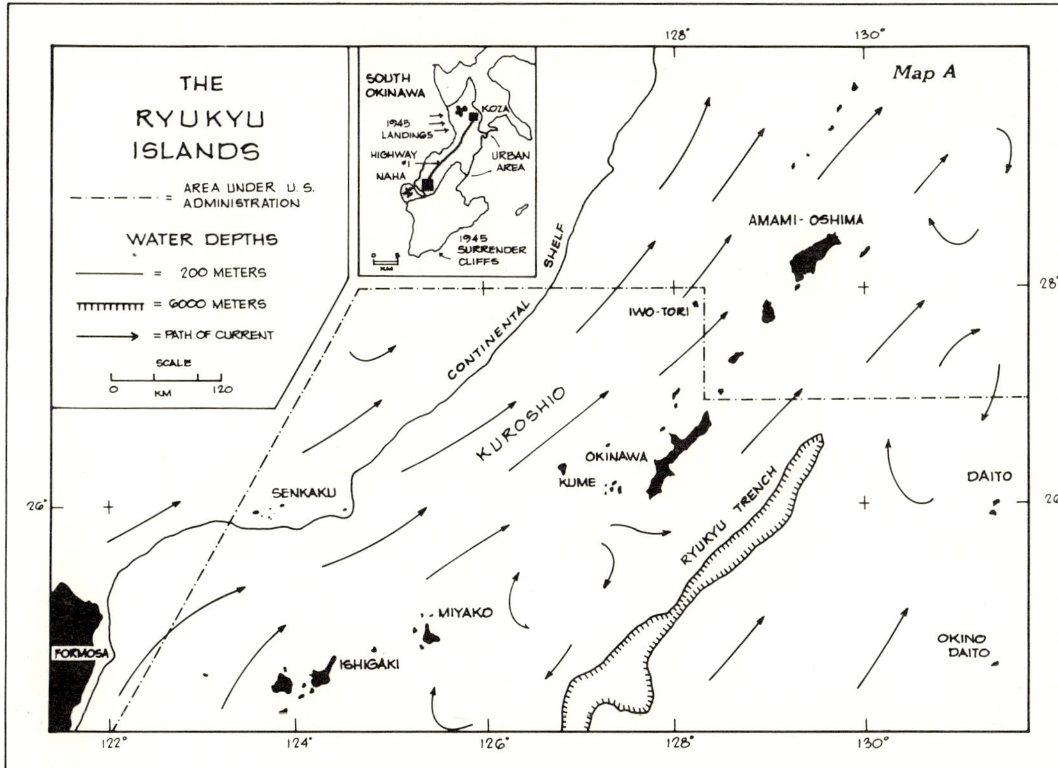
Map A
THE
RYUKYU
ISLANDS
= AREA UNDER U.S. ADMINISTRATION
WATER DEPTHS
= 200 METERS
= 6000 METERS
= PATH OF CURRENT
SCALE
0
KM
120
SOUTH OKINAWA
1945 LANDINGS
HIGHWAY #1
NAHA
KOZA
URBAN AREA
1945 SURRENDER CLIFFS
KM
CONTINENTAL SHELF
KUROSHIO
RYUKYU TRENCH
FORMOSA
SENKAKU
ISHIGAKI
MIYAKO
KUME
OKINAWA
IWO-TORI
AMAMI-OSHIMA
DAITO
OKINO DAITO
122°
124°
126°
128°
130°
26°
28°

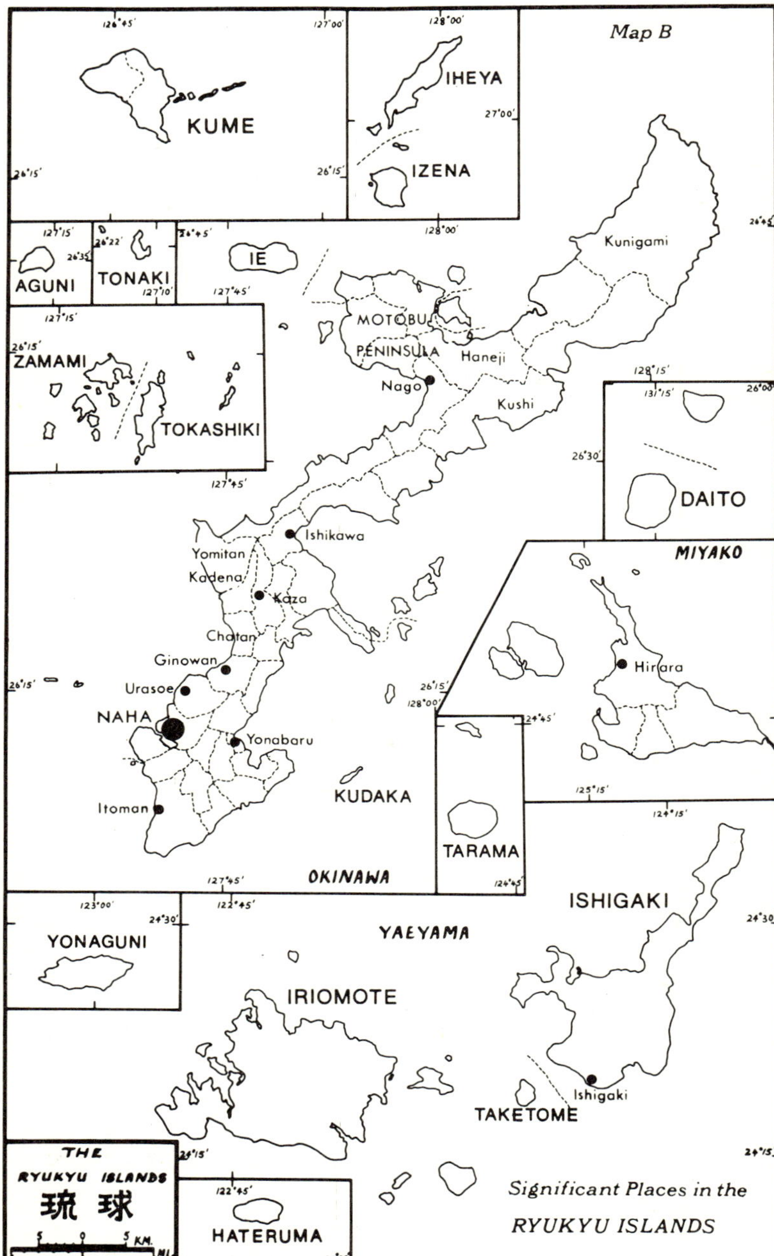
Map B
KUME
IHEYA
IZENA
AGUNI
TONAKI
IE
ZAMAMI
TOKASHIKI
MOTOBU PENINSULA
Kunigami
Haneji
Nago
Kushi
DAITO
MIYAKO
Hirara
Ishikawa
Yomitan
Kadena
Koza
Chatan
Ginowan
Urasoe
NAHA
Yonabaru
Itoman
KUDAKA
TARAMA
OKINAWA
YONAGUNI
YAEYAMA
ISHIGAKI
IRIOMOTE
Ishigaki
TAKETOME
HATERUMA
THE RYUKYU ISLANDS
琉球
KM.
ML.
Significant Places in the
RYUKYU ISLANDS

panded by the American military forces. Koza, whose name was changed in 1974 to Okinawa City, had a population of 58,658 in 1970. In contrast to this urbanized area are the relatively sparsely settled areas in northern Okinawa and in the southern islands. Iriomote Island, for example, located in the Yaeyama group of islands is twice the size of the Naha-Koza urban area but had a population of only 2,300 persons in 1970. It has now become a National Park of Japan.

The Ryukyu Islands comprise seventy-three distinctive islands plus innumerable coral reefs and offshore rocks. Map A shows the location and the general bathymetry around the islands with the deep Ryukyu Trench to the east and the Asian Continental Shelf to the west. Okinawa is the largest island, some 60 miles in length and varying from 2 to 14 miles in width. Forty-three islands are inhabited though some have only a handful of people. In 1970, the Ryukyu Islands were subdivided into fifty-nine political administrative units, classed as cities (*shi*), towns (*cho*), and districts (*son*). The cities and towns and some of the districts and islands are named on Map B.

The Ryukyu Islands have limited natural resources, though they have considerable diversity in their physical geography. Much of the land is hilly and mountainous and cannot be used for agriculture. Only 21.5 per cent of the total area of 2,239 sq. km. (865 sq. mi.) was cultivated in 1971. Sweet potatoes, vegetables, and small amounts of rice are grown, and limited amounts of fish are caught in coastal waters for food consumption by the local population. Of commercial importance are sugarcane, the major crop, and pineapples grown and processed for price-subsidized markets in Japan.

The humid subtropical climate of the Ryukyu Islands is moderated by the warm Kuroshio Current, flowing from the south. The January and February temperatures at Naha average 71°F, and the July temperatures average eleven degrees warmer, 82.2°F. The rainy season in the late spring and early summer is influenced by the Asian monsoon air drifts. The average annual rainfall at Naha is 84.5 in., one-fourth of this occurring in the months of May and June. Four or five severe typhoons can be expected annually to affect some parts of the Ryukyu Islands in the late summer and early fall. Crops, houses, stone fences, roads, and sea-walls, as well as all human activities, are influenced by and must be adapted to survive violent typhoon winds, heavy rainfall, and salt sea spray.

2. Traditional Agriculture

TRADITIONAL agriculture in the days of the Liu Chiu Kingdom was characterized by intensive use of the limited arable land for growing food for local consumption. The Liu Chiu Kingdom had its origin in the distant past. Three small principalities or chieftainships were finally unified in 1429 under the Sho dynasty. This dynasty lasted until 1879 when the Sho king was deposed by the Japanese and Okinawa was merged into Japan.[4] The predominant activity of the people throughout the history of the Liu Chiu Kingdom was agriculture, although during its Golden Age in the fifteenth and sixteenth centuries the Liu Chiu Kingdom benefited from an extensive foreign trade, capitalizing upon its entrepôt position in the Far East. After the kingdom was subjugated by the feudal lords of Satsuma in 1609, sugar became an agricultural export product and was used to pay the taxes demanded by Japan.

In the farming system as practiced during the history of the Liu Chiu Kingdom much hand labor was involved. All able-bodied members of the farm households worked in the fields. Most of the farming people lived in villages, going out to till their fields. Some, however, were living in dispersed settlements, particularly where lands had been newly reclaimed. Within the agricultural villages the small plots of cultivated land were periodically reassigned to the families who made up a village.[5] It has been estimated that this communal system comprised roughly 70 per cent of the cultivated land, with the other 30 per cent owned by individual families who had reclaimed the lands themselves.[6]

The introduction of the sweet potato from China in 1605 made

4. A very readable history of the Ryukyus is George H. Kerr, *Okinawa, The History of an Island People* (Rutland, Vermont, and Tokyo, Japan: Charles E. Tuttle Co., 1958).

5. There is considerable literature on the communal villages. One useful source, Hiroshi Tamura's *Ryukyu Kyosan Sonraku no Kenkyu* [A Study of Ryukyuan Communal Villages], Tokyo: Oka Shoin, 1927, was translated by Shogo Oiyama in 1955, and xeroxed copies of the 467-page translation are available in some library and research collections. A recent English-language study of this land tenure system, called *jiwari seido,* is that of Kenji Kenneth Oshira, *Jiwari Seido in Central and Southern Ryukyus* (Master's thesis, University of Washington, 1965).

6. This estimate is given in F. R. Pitts, W. P. Lebra, and W. P. Shuttles, "Post-war Okinawa," mimeographed (Washington, D.C.: Pacific Science Board, National Research Council, June 1955).

possible a considerable increase in cultivated land, since sweet potatoes could be grown on steep slopes and were not greatly harmed by salt sea spray or typhoon winds. Sweet potatoes provided a staple food for the villagers and the surplus tubers and plentiful greens were excellent fodder for pigs. As sweet potato production grew and spread to the outer islands and to northern Okinawa, the population increased. Rice was a valued food crop, but only limited areas of alluvial land were suitable for the development of paddy fields. Millet, other cereals, and various kinds of vegetables were also grown under the subsistence agricultural system which prevailed for many centuries.

The increase and subsequent decrease in the dry fields and the paddy fields on Okinawa Island from 1611 through 1971 is shown graphically in Chart 1. Unfortunately data are not available for the southern Ryukyu Islands in the period before 1904. There was relatively little development of those areas for other than subsistence crops during the period of the Liu Chiu Kingdom.

Interesting accounts of agriculture on Okinawa Island were written by members of the American Expedition under Commodore Matthew C. Perry which visited Naha in 1853–54. This expedition opened Japan to the outside world, and between visits to Japan, the ships wintered in the waters off Naha. One of the members of the expedition was James Morrow who was assigned the task of bringing back seeds and plants to the United States which might be useful in American agriculture. Morrow's *Report* is in manuscript form in the National Archives and has been reprinted in various publications.[7]

Morrow's first view approaching Naha was ". . . beautiful. From the beach to the summit of the long and gently ascending slope, nearly the whole surface seems to be in the highest state of cultivation. The shades of green presented by the different winter crops; the uniform, terraced hills; the occasional patches of large, spread-

7. The *Research and Information Papers* of the Ryukyu Island Project reprinted accounts of four scientists who accompanied Commodore Matthew C. Perry on the expedition to Japan in 1853–54. These papers are listed in Appendix A and comprise numbers 12 through 15. Included in number 12 is an appendix giving a bibliography on readily available material on the expedition. The reports of these scientists are in volume 2 of *Narrative of the Expedition of An American Squadron . . .* by Francis L. Hawks (Washington, 1856). The best account of Dr. James Morrow's life is in the editor's introduction to Allen B. Cole, ed., *A Scientist with Perry in Japan, The Journal of Dr. James Morrow* (Chapel Hill: University of North Carolina Press, 1947).

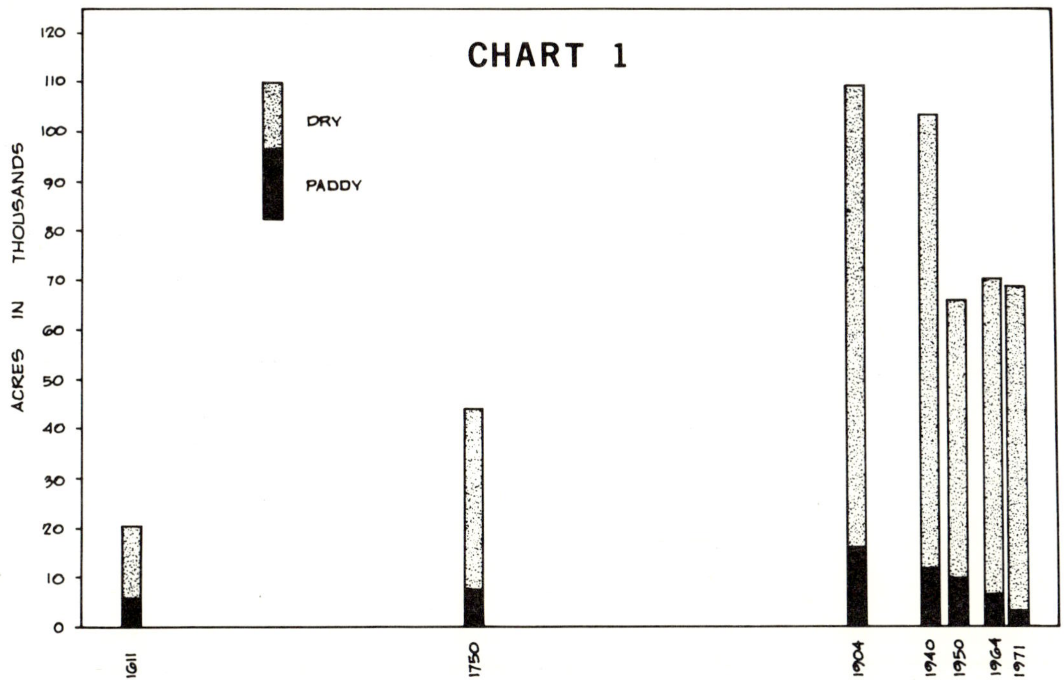

CULTIVATED LAND IN OKINAWA ISLAND

ing forest-trees, forming beautiful groves to relieve the uniformity of cultivated fields; all conspire to form a most rich and highly cultivated rural landscape. . . . Every foot of ground appears to be carefully cultivated, unless from situation or barrenness cultivation would be unprofitable. Patches of sweet potatoes meet the eye in every direction. . . . In the neatness of agricultural operations the Lew-Chewans are seldom surpassed. . . . It would be difficult to introduce improved agricultural implements among a people where labor is so cheap, and where any cheapening of agricultural operations would be so little advantage to the laboring classes. . . . At present indications are that the population is increasing, which is a great evil when the soil has already attained its maximum, and where there is no outlet for the population by emigration."

Another member of Perry's expedition was Fleet Surgeon Daniel S. Green. Along with Morrow and others he took a five-day trip from Naha to Shioya Bay in northern Okinawa. He was particularly impressed with the care taken in terracing fields. As he noted, "Two objects seem to influence the islanders in all their operations, viz: the retention of water or moisture, and the avoidance of surface-washing." He wrote that since the land being cultivated "was divided into very small portions or parcels, and the labor being done by man, *horticulture* would seem to be a more appropriate term than *agriculture,* and the whole country to be overspread by a series of gardens rather than farms." He concluded his report in which he detailed many aspects of the crops and their production with these words: "The present system of agriculture could scarcely be improved for the Lew-Chewans, considering the circumstances in which they are placed."

The Liu Chiu Kingdom at the time of Perry's visit in 1853–54 had a population of an estimated 300,000 persons. The majority were engaged in agriculture and handicraft industries. Sugar was the major commercial crop by which most of their heavy taxes were paid. Sugarcane had been initially introduced into the Ryukyu Islands in 1374, and it became much more important after 1623 when better methods of extraction of the sugar were adopted. Exports of sugar increased considerably. However, the profits were realized chiefly by the Liu Chiu court, the Satsuma feudal lords who had conquered the Ryukyus in 1609, and the Kagoshima and Osaka sugar merchants, leaving little gain to the Liu Chiu farmer. After the political annexation of Okinawa to Japan in 1879, the

Japanese built modern sugar mills and introduced better types of sugarcane. Sugar continued to be the leading commercial crop.[8]

With a complex geologic structure and with varied climates and soils, regional differences in agriculture developed in the Ryukyu Islands. Being an archipelago, the numerous islands developed differently depending on their size and distance from the main island of Okinawa. By 1903, village communal holdings had been converted to permanent individual holdings with the exception of those on small Kudaka Island where they remained for some time as interesting relics.[9] The data available for 1904 reveal that Okinawa Island was relatively intensively cultivated; 31 per cent of its total area was under cultivation at that time. By comparison the malaria-ridden and typhoon-prone southern and outlying islands of Yaeyama and Miyako were little developed at that time. The farmers on those islands and in northern Okinawa continued to live on a subsistence basis with limited access to the outside world and to Japanese markets. In periods of famine they had to resort to cycads for food.

Because of the Japanese desire for self-sufficiency in sugar, production of cane sugar was fostered on the limestone-based soils of southern Okinawa, Miyako, and the Daito Islands in the period before World War II. Much of the sugarcane was ground for syrup in simple horse-powered stone mills, then boiled in large iron kettles heated by wood fires. Most of the raw brown sugar (often called

8. A detailed discussion, based on a meticulous survey of manuscript and other materials, of the early exports of sugar may be found in Mitsugu Sakihara, *The Significance of Ryukyu in Satsuma Finances during the Tokugawa Period* (Ph.D. diss., University of Hawaii, August 1971). His chapter "Satsuma's Monopoly of Ryukyu's Native Products," pp. 110–39, is especially useful. Tomotaka Ishimine's Ph.D. dissertation, *Japanese Tariff Preference and the Okinawan Sugar Industry* (Ann Arbor, Mich.: University Microfilms, 1972) gives a great deal of useful information, largely drawn from Japanese sources, on the relations of Japanese companies and Okinawan sugar production, processing, and trade. A Japanese official, Kitokuro Ichiki, traveled widely and observed conditions in Okinawa Prefecture in the early 1890s. His report, *Ichiki Shokikan Torishirabe Sho* [Official Report of Investigation and Study by Government Clerk Ichiki] was translated by Shogo Oiyama for the United States Civil Administration of the Ryukyu Islands in 1956. This report gives some interesting details on the conditions in agriculture and, in particular, sugar production at the time of his survey.

9. Richard W. Lieban completed his Ph.D. dissertation on *Land and Labor on Kudaka Island* at Columbia University in 1956 (Ann Arbor, Mich.: University Microfilms). He wrote a short and interesting article, "The Land Systems of Kudaka Island," *Sociologus* 15 (1955): 150–56.

black sugar) was exported to markets in the cities of Japan. Table 1 gives data on the production of centrifugal and noncentrifugal sugar in Okinawa Prefecture from 1918 through 1940.

TABLE 1
PRE–WORLD WAR II SUGAR PRODUCTION IN OKINAWA PREFECTURE, 1918–40
(in metric tons)

Year	Centrifugal	Non-centrifugal	Total
1918	13,419	48,084	61,503
1919	18,868	29,200	48,068
1920	16,748	48,212	64,960
1921	14,325	39,463	53,788
1922	16,310	30,384	46,694
1923	21,791	38,421	60,212
1924	13,926	31,610	45,536
1925	19,763	46,823	66,586
1926	18,873	45,590	64,463
1927	18,875	45,417	64,292
1928	23,088	49,128	72,216
1929	21,232	53,905	75,137
1930	20,900	40,610	61,510
1931	20,204	46,103	66,307
1932	23,899	62,634	86,533
1933	23,690	63,871	87,561
1934	21,852	53,456	75,308
1935	22,106	65,219	87,325
1936	25,809	74,306	100,115
1937	23,778	54,471	78,249
1938	28,631	57,161	85,792
1939	35,298	86,899	122,197
1940	30,241	65,086	95,327

SOURCES: Tomotaka Ishimine, *Japanese Tariff Preference and the Okinawan Sugar Industry* (Ph.D. diss., University of Wisconsin, 1971; Ann Arbor, Mich.: University Microfilms, 1972), Table 6, p. 57. The table was drawn from Japan Sugar Stabilization Board, *Nippon no Togyo-shi* [History of the Japanese Sugar Industry] (Tokyo, 1967).

3. Pre–World War II Agricultural Situation

IN 1937 Japan embarked on its disastrous course of war with China. That same year marked an important turning point in the evolution of the economy of the Ryukyu Islands. Relatively few Japanese troops were stationed in Okinawa at that time; it was not until 1944 that large Japanese military forces were deployed there. In 1937, lured by the inviting prospects of economic development in Taiwan, in Southeast Asia, and hopefully in China, Japan continued to neglect Okinawa. It remained the poorest prefecture of Japan with the resultant out-migration of a considerable number of Okinawans. In the period 1936–38, 9,670 persons emigrated, including 1,245 who went to Brazil and 5,313 to the Philippines, mainly to Davao where Japanese were developing the abaca industry in

TABLE 2
ACREAGE OF CULTIVATED LAND, OKINAWA PREFECTURE, 1940

		Okinawa	Miyako	Yaeyama	Total
Owned land	Dry	75,931	28,313	7,647	111,891
	Paddy	9,853	336	3,711	13,900
Tenant land	Dry	16,343	1,805	2,101	20,249
	Paddy	1,675	30	677	2,382
Total land	Dry	92,274	30,118	9,748	132,140
	Paddy	11,528	366	4,388	16,282

SOURCE: *1969 Okinawa Statistical Yearbook* (Naha: Agency of Statistics, Planning Department, Government of the Ryukyu Islands, 1971), Table 1, p. 375. Data given in *tan* and converted to acres (1 *tan* = 0.245 acres).

Mindanao. In the same period, 1,037 Okinawans migrated to Hawaii, the last appreciable number to emigrate there.[10]

Since the pre-World War II period was a critical time for the economy, it is worthwhile considering briefly the agriculture of Okinawa Prefecture of that period. The total cultivated land for all of Okinawa Prefecture in 1940 was 148,422 acres. This acreage was only slightly higher than the amount of land cultivated in 1937. Table 2 gives data on the ownership and type of cultivated acreage for 1940. It is obvious from the table that most of the cultivated land was on the island of Okinawa. Miyako, underlain with limestone rock formations, had very few paddy fields because of the

10. The data on emigration are from a table in *1969 Okinawa Statistical Yearbook* (Naha: Agency of Statistics, Planning Department, Government of the Ryukyu Islands, 1971), p. 376.

character of its terrain and soils. Only a small percentage of the total area of Yaeyama was cultivated land, for development progressed slowly there due to its isolation and malarial infestation. Eighty-five per cent of the land cultivated by farm households in Okinawa Prefecture was owned by those who tilled it. Only 15 per cent of the cultivated land was leased by tenants.

The acreage of the cultivated land devoted to the leading crops in Okinawa Prefecture in 1937 is given in Table 3. Many of the paddy fields, particularly on the island of Okinawa, were double-cropped with rice or with rice and some winter crops. Sugarcane was raised on an approximate 18-month cycle of growth; many sugarcane fields were not freshly planted, but were ratooned from roots left in the fields. Minor crops and vegetables grown in gardens near the farmhouses were not included in the data. Table 3 indicates the importance of sweet potatoes. They occupied roughly half of the acreage of land in crops in 1937. The limited development of Yaeyama by 1937 is very apparent. Though Yaeyama was roughly 26 per cent of the total land area of Okinawa Prefecture, at that time it had only 9 per cent of the cultivated land.

TABLE 3

ACREAGE IN MAJOR CROPS, OKINAWA PREFECTURE, 1937

		Okinawa	Miyako	Yaeyama	Total
Rice	1st crop	9,664	323	4,395	14,382
	2d crop	6,180	79	1,189	7,448
	Total	15,844	402	5,584	21,830
Wheat		948	1,910	550	3,408
Soybeans		9,880	1,894	333	12,107
Sweet potatoes		57,411	10,591	6,326	74,328
Sugarcane[a]		31,368	5,097	478	36,943
Total		115,451	19,894	13,271	148,616

SOURCE: *1969 Okinawa Statistical Yearbook,* Table 2, p. 375. Data converted from *tan* to acres (1 *tan* = 0.245 acres).

a. Data for sugarcane acreage are from 1936 since most of the harvesting took place in 1937.

There was only one major urban area in 1937, the Naha-Shuri district. Itoman, south of Naha, had a relatively large number of families engaged in fishing. Otherwise, the number of farm households was in close relation to the number of inhabitants in each

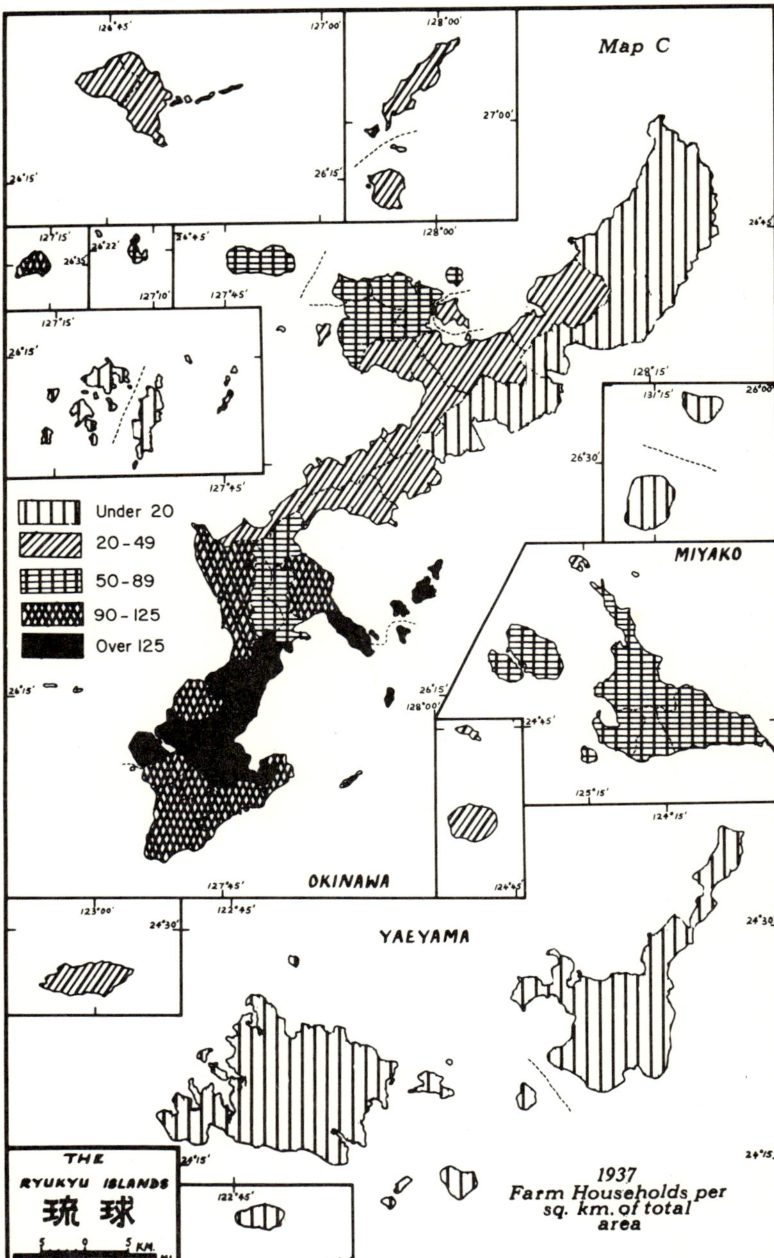

Map C
Under 20
20-49
50-89
90-125
Over 125
OKINAWA
MIYAKO
YAEYAMA
THE
RYUKYU ISLANDS
琉球
KM.
ML
1937
Farm Households per
sq. km. of total
area

political subdivision. Naturally there was a considerable geographical difference in the number of farm households per square kilometer of total area. This distribution is shown on Map C. In this study there is a series of maps—C, D, E, and M—which shows the distribution of farm households per square kilometer of total area for each of the fifty-nine political subdivisions for the years 1937, 1951, 1964, and 1971. These maps may be compared in order to show the geographical aspects of the changes which have taken place in the agriculture of the Ryukyu Islands. By relating the number of farm households to total area rather than to cultivated area, the changing ratio of farmers to land may be shown effectively, since the land factor is kept constant.[11]

In 1937 there were 38 farm households per square kilometer of total area for all of Okinawa Prefecture (with the exception of the Daito Islands for which data are not available but which were probably well below the average). There was, however, a wide range, from over 100 farm households per square kilometer in the lowlands around the Naha-Shuri area to 3.7 households per square kilometer in Taketomi-son, a district which includes mountainous Iriomote Island in Yaeyama. In relation to the total area the mountainous northern part of the island of Okinawa also had few farm households. Some of the offshore islands, Aguni, for example, had a relatively large number of farm households per square kilometer of total area. The lowest densities, less than 10 farm households per square kilometer of total land, are in northern Okinawa and in Taketomi-son. The highest densities, over 90, are in the area of south-central Okinawa.

4. Agricultural Situation, 1945–52

THE effects of World War II were very great on the Ryukyu Islands and will likely be evident in one form or another for decades to come. As the tides of war were turning against Japan in 1944, the Japanese High Command's decision to fortify Okinawa

11. Appendix B gives a brief description with some illustrations of a computer-drawn map of the Ryukyu Islands prepared as part of the Ryukyu Islands Project by Steven T. Gladin.

made it a magnet attracting American and Allied forces bent on invading the homeland of Japan. The Battle of Okinawa in the spring of 1945 devastated southern Okinawa. It has been estimated that as many as 100,000 Okinawan civilians were killed during the ninety-day battle.[12] This sacrifice of the people of Okinawa for the defense of Japan will never be fully expiated.

An aftermath of the Battle of Okinawa was the establishment of the American military and civil administration of the Ryukyu Islands from 1945 to 1972. After World War II an estimated 150,000 persons were repatriated to the Ryukyu Islands. This included some Okinawans who had earlier migrated to Japanese-held areas in Southeast Asia and in the Pacific Islands. It also included some of those who had enlisted or had been drafted into the Japanese military forces. To help feed the populace, relief foods were imported from the United States. The Okinawan farmers of necessity grew food crops; the export of sugar was limited and little was produced. Much former agricultural land had been taken over by the Japanese military forces for airfields and other installations as Okinawa was fortified. Still more farmland was taken over for additional military bases which were built in the Ryukyu Islands by the United States Armed Forces. Thus, agriculture in the Ryukyu Islands was at a low point in 1949.

The situation in this period was well summarized in a report prepared by a group of agricultural experts who studied the situation in 1949. This mission was headed by Raymond E. Vickery and the report is usually called by his name.[13] The study done by the Agricultural Mission made numerous recommendations, many of which were followed. The first paragraph of the Vickery Report is worthy of quotation: "Four years after the cessation of combat operations, the people of the Ryukyu Islands and especially Okinawa, remain in a state of abject poverty and a continuing burden on the American taxpayer. Little prospect exists for their attaining a significantly greater degree of self-sufficiency until an effective

12. The number of casualties must of necessity be estimates. The best estimates were made by Col. Norman D. King whose memorandum on the subject was reprinted in *Research and Information Paper*, Ryukyu Islands Project, number 19 (November 1972). This paper also includes a bibliography on the Battle of Okinawa.

13. The Vickery Report is entitled "Agriculture and Economic Reconstruction in the Ryukus, A Report," mimeographed (Washington, D.C.: Agricultural Mission to the Ryukus, November 1949).

program of reconstruction is inaugurated and vigorously pursued."

After the start of the Korean War in 1950 the Ryukyu Islands grew in importance as an American military supply and staging base. Following some of the recommendations of the Vickery Report of 1949, the United States embarked on a policy of fostering the agricultural development of the islands. Considerable amounts of land around the military bases were released for agricultural uses. Crop production began to rise. The United States Civil Administration, formed in 1950, and the Government of the Ryukyu Islands, organized in 1952, initiated programs to develop the agricultural and human resources. Numerous research studies of the situation were undertaken.[14] A population census was taken on October 1, 1950 and an agricultural census for 1950 was taken on February 1, 1951. Thus, the years of 1950 and 1951 are significant in the analysis of the geographical aspects of agricultural change in the Ryukyu Islands.

Reconstruction of agriculture was indeed greatly needed. As the Vickery Report noted: "The war, and particularly the campaign for Okinawa had a devastating effect on the Ryukyuan economy. Food production was prostrated. Farm buildings, livestock and equipment, as well as fishing fleets and industrial facilities, were almost completely destroyed on Okinawa. The military used farm land arbitrarily with no regard to its economic value. . . . Post-war reconstruction has been extremely slow. It has been hampered by lack of adequate policies, insufficient emphasis on the revival of food production and increased population pressure. The simple fact that the all-important agriculture and fishing industries must be rebuilt for the recovery of the Ryukyuan economy has been ignored. Okinawa has remained in a particularly deplorable state. Nearly one-fourth of the arable land continues to be used for military units."[15]

The report was very critical of the "one-mile rule" which prohibited any Okinawan farmers from having buildings within one mile of any United States billets. This meant, according to the

14. Many of these studies carried on in the early 1950s were supported by grants administered by the Pacific Science Board of the National Research Council. As a part of this research project, each of the areas was visited in 1970 and up-to-date maps of land use prepared for those areas which had been previously mapped. Appendix C lists some of these reports and includes a map locating the areas where field studies were made.

15. Vickery Report, p. 1.

report, that: "Tools and supplies must be carried farther, the farmers must walk farther to tend the fields, and the crops must be hauled farther for use or storage. . . . the boundaries of military installations have not been stabilized, but rather change continuously so that farmers never knew what land will be available in the future. A cursory observation of the country-side reveals that to date no thought has been given to efficient use of the land. In such a completely agrarian economy this has had a devastating effect."[16]

TABLE 4

ESTIMATED ACREAGE OF MAJOR CROPS, RYUKYU ISLANDS, 1949

Island Group	Okinawa	Miyako	Yaeyama	Total
Rice, 2 crops	14,000	300	5,700	20,000
Small grains and millet	2,500	6,000	900	9,400
Soybeans and peanuts	8,400	4,000	900	13,300
Sweet potatoes	58,000	16,000	6,760	80,760
Sugarcane	1,000	3,000	50	4,050
White potatoes and vegetables	10,600	1,250	350	12,200

SOURCE: "Agriculture and Economic Reconstruction in the Ryukyus: A Report," mimeographed (Washington, D.C.: Agricultural Mission to the Ryukyus, November 1949). Data drawn from tables in the appendix. The report notes the fact that agricultural data were not gathered by a central crop reporting service and were very crude estimates.

Table 4 gives estimates of the acreages for the major crops in 1949 according to the Vickery Report. These estimates appear generally to be derived from prewar data, adjusted for Okinawa to take account of the devastating effect of the Battle of Okinawa and the subsequent disruption of the agrarian economy.

Because of the context and the time in which it was written the report gave considerable emphasis to the need for rice production: "Travel through the Northern, Central and Southern Ryukyus clearly indicated that rice is grown on all possible land where the topography is suitable and water supplies are adequate. There seems to be no question that rice is preferred on all land suited for its production. Rice is quite commonly grown as upland rice, de-

16. Ibid., p. 22.

pending on rainfall only as the water supply. In general this rice is poor, and in certain seasons may be near failure."[17]

In regard to sugarcane the report summarized the advantages and disadvantages for sugarcane production in the Ryukyu Islands and reached this conclusion: "At the present time, it would neither seem advisable to encourage or discourage the production of sugar cane, pending the outcome of policies regarding free trade and the economic conditions governing the world price of sugar." The report went on to recommend that "special trade agreements, however, should be worked out with Japan so that such quantities of sugar as continue to be produced in the Ryukyus can be marketed to help pay for needed Japanese imports."[18]

Because of the repatriation of large numbers of people, the total number (82,817) of farm households which cultivated more than 5 *ares* (0.1235 acres) of land in 1951 was not much less than that for 1937. There were 37 farm households per square kilometer of total area for the Ryukyu Islands; this was only a slight reduction from the 38 farm households per square kilometer in 1937. The geographical distribution and densities did vary, however, from what they had been in 1937. For example, the "one-mile limit" drastically lowered the number of farm households around the Kadena Air Base. Map D gives the number of farm households per square kilometer of total area in 1951 by political subdivisions. Sparse settlement patterns still continued in northern Okinawa, in the Daito Islands, and in Yaeyama as had been the case in 1937. Moreover, in 1951, there were relatively few farm households either in the areas devastated during the Battle of Okinawa or around the military bases. The urban area of Naha in the October 1950 census had 108,662 residents out of a total population of 698,827 for the Ryukyu Islands; this concentration of urban population indicates that some economic recovery had taken place.

As already noted, the period 1950–52 marked a turning point in agricultural development of the Ryukyu Islands. The Korean War led to a substantial increase of American military installations and expenditures in Okinawa. Various factors combined to foster agricultural development. The United States Civil Administration and the Government of the Ryukyu Islands held stable government policies and encouraged agriculture in many ways. The signing of the

17. Ibid., p. 49.
18. Ibid., p. 56.

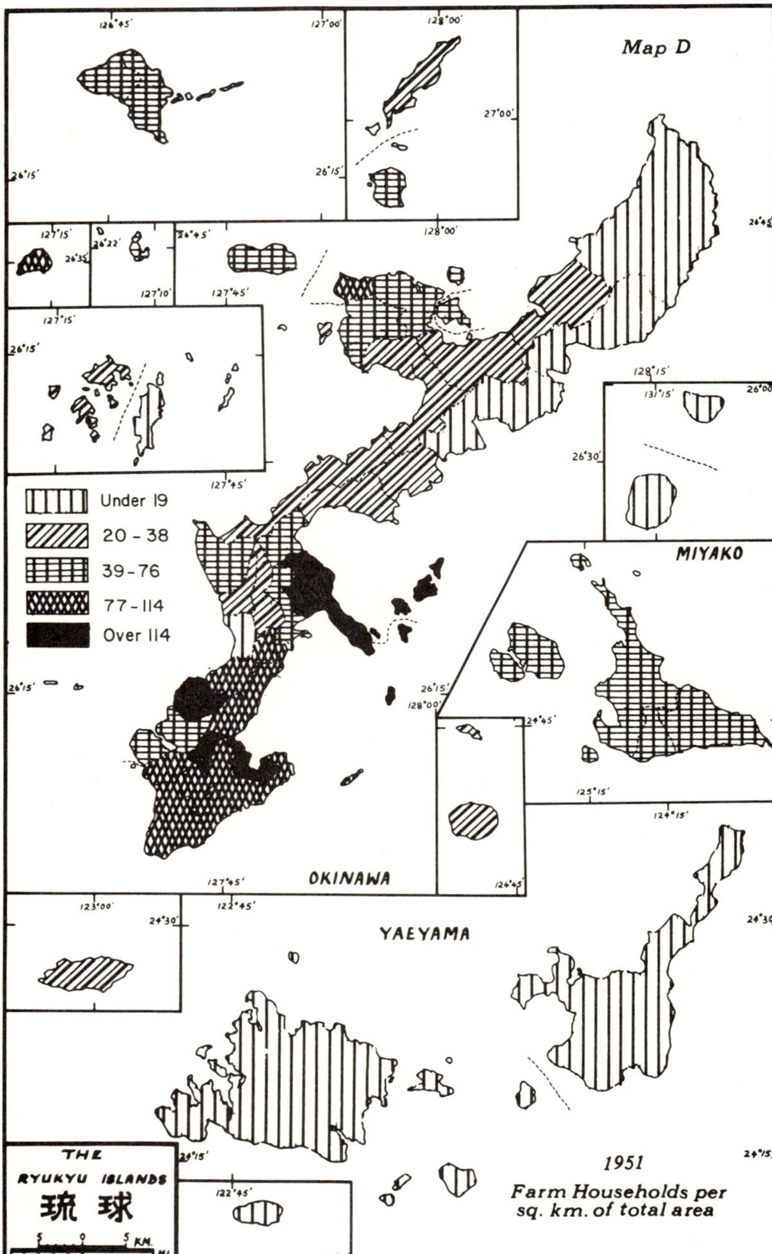

Map D
Under 19
20 - 38
39-76
77-114
Over 114
OKINAWA
MIYAKO
YAEYAMA
THE
RYUKYU ISLANDS
琉 球
5 0 5 KM.
1951
Farm Households per
sq. km. of total area

Peace Treaty with Japan provided for the continuation of the United States administration of the Ryukyu Islands. There was a growing awareness by the United States that, as it continued its military presence in the Ryukyus, much more attention needed to be directed to the agrarian sector and to the well-being of the Ryukyuan people. The "one-mile limit" was abrogated. Military bases were stabilized and some of them considerably contracted in size. In some military bases arrangements were made so that Okinawan farmers could lease and cultivate agricultural land in the periphery of the military installations and under telecommunication wires.

The very knotty problems of compensation to the former owners for the land which was used for military bases on Okinawa were gradually worked out, though not without considerable agitation and turmoil. The recrimination and ill-will engendered among resentful farmers was slow in subsiding. The problems were very difficult to solve. Many land records had been destroyed during the Battle of Okinawa. Customarily the land had been divided into many small plots. The setting up of a joint American-Ryukyuan Land Tribunal and its effective operation helped to alleviate the situation. Eventually Ryukyuan land owners were paid rental for their land; even the pre-Peace Treaty claims were met and rental payments made for the period 1945–52. These actions took many years to work out, however.[19]

Closely following on these American actions, the Japanese government instituted policies for the subsidization of prices for sugar and pineapples from the Ryukyu Islands. In part, the sugar price-subsidy program was a result of the uncertainty of worldwide supplies. Another factor, however, was the program's usefulness politically as a way of influencing the people of the islands.[20]

19. There is a rather vast amount of literature on the land claims problems in the Ryukyu Islands. One of the best brief accounts which presents the different sides objectively is in "The Military Base Problem," pp. 40–56, in Mikio Higa's *Politics and Parties in Postwar Okinawa* (Vancouver: University of British Columbia, 1963). Views on the United States' position and action may be noted in the various *Annual Reports* of the United States Civil Administration of the Ryukyu Islands. The Government of the Ryukyu Islands prepared a document, *Military Land Problems in Okinawa* (Naha, June 1958), for its delegation to the United States for negotiations.

20. The dissertation of Tomotaka Ishimine, *Japanese Tariff Preference and the Okinawan Sugar Industry,* places emphasis on the economic and business factors rather than the political factors, but it is of value.

Sweet potato fields in the foreground; paddy fields, sugarcane, and mountain slopes in the background.

The rolling fields devoted to sugarcane and vegetables in Kochinda near Naha in southern Okinawa.

Sugarcane is the predominant land use in the agricultural landscape of southern Okinawa. The village in the background is Hanashiro.

Agricultural landscape in Ishigaki with limited cultivated land.

5. Agricultural Situation, 1964

BY 1964, the various agricultural development programs, plus the diligent efforts of many Ryukyuan farmers, had achieved dramatic results in improving the agricultural situation in the Ryukyu Islands. The Japanese government's sugar liberalization policy classed Okinawa as "domestic" in terms of sugar production and sugar sales to Japan. The policies which had been operating since 1952 had provided price subsidies for sugar from the Ryukyus and the sugar liberalization program stabilized these subsidies. Table 5, drawn from Ryukyu Sugar Industry Association reports and newspaper annuals by Tomotaka Ishimine, gives data on the sugar exports from the Ryukyu Islands in 1952–65. The exports in 1952, which included no centrifugal sugar and only 5,465 tons of noncentrifugal (or "black") sugar, contrasted greatly with the 1965 exports of 268,000 tons of centrifugal sugar and 23,402 tons of noncentrifugal sugar. These figures are dramatic indications of the tremendous impact that Japanese subsidization programs had on production and marketing of sugar in the Ryukyu Islands. As Tomotaka Ishimine notes in his dissertation: ". . . the Okinawan production cost exceeds the Japanese import price by a considerable margin. For example, the production cost in 1964–1965 was $188.80 per ton while the Japanese C.I.F. price for the comparable year was $69.60 per ton. Only by means of the high duty of $115.30 per ton and the subsidy through the purchase price of $10.00 in addition to the duty exemption was Okinawan sugar made competitive in the import market of Japan."[21]

The benefit of the price subsidization by the Japanese govern-

21. Ibid., p. 19.

TABLE 5
EXPORTS OF SUGAR, RYUKYU ISLANDS, 1952–65

	Centrifugal Sugar		Noncentrifugal Sugar	
Year	Metric Tons	Value in $ thousands	Metric Tons	Value in $ thousands
1952	0	0	5,465	1,326
1953	0	0	7,623	1,642
1954	451	113	18,783	4,380
1955	6,685	1,427	24,744	5,538
1956	5,745	1,072	28,639	5,463
1957	2,554	555	28,527	5,862
1958	1,950	355	38,267	7,234
1959	10,740	2,541	30,128	5,713
1960	27,439	7,084	28,139	5,680
1961	58,572	14,354	14,825	3,188
1962	92,567	22,792	23,743	4,762
1963	140,693	39,919	17,393	4,432
1964	143,021	39,579	11,417	2,761
1965	268,000	49,576	23,402	3,691

SOURCES: Tomotaka Ishimine, *Japanese Tariff Preference and the Okinawan Sugar Industry,* Table 5, p. 54. The data are drawn from the Ryukyu Sugar Industry Association, *Togyo Kankei Tokei Shiryo,* and the Okinawa Times, *Okinawa Nenkan.*

ment went largely to the Ryukyuan farmers. They had responded as would be expected during the previous decade by greatly expanding the acreage of sugarcane fields. Better varieties of sugarcane, notably NC–310, were introduced. Better tillage practices were instituted giving increased production per acre. New centrifugal sugar mills were constructed and the price of sugar became a highly political matter. After analysis of the data, including data from reports of the Government of the Ryukyu Islands, Tomotaka Ishimine concludes that: "Since 87.5 percent of the total farm families, or approximately 63,356 of the 77,500 farm families, engaged more or less in cane production, the average subsidy per family was $145.50 in 1965."[22]

In the spring of 1964 an agricultural census was taken in the Ryukyu Islands. This was a very detailed census with a great mass of material collected on every phase of agriculture. The results of this census were published subsequently in a series of volumes in

22. Ibid., pp. 90, 91. He cites as a source a Government of the Ryukyu Islands report, *Okinawa Tohyo no Genkyo* [The Present Situation of the Okinawan Sugar Industry].

the Japanese language. The summary tables were reproduced bilingually in the annual *Statistical Yearbooks* of the Government of the Ryukyu Islands. This agricultural census provided significant data for government officials, research scholars, and the general public.

One of the important tabulations made in the 1964 Agricultural Census was that of the land used by farm households for purposes other than crop production. This land amounted to 48,755 acres, an appreciable amount. This included 21,471 acres in forest land (including wood lots); 4,876 acres which had once been cultivated but which was out of crop production at the time of the census; 13,446 acres of grassland which had never been cultivated; 6,902 acres used for building sites by the farm families; and 2,060 acres used for miscellaneous purposes such as grave sites.

The land cultivated for crops by the farm households in the Ryukyu Islands totaled 122,158 acres in 1964. To this should be added a small amount of land, 2,885 acres, cultivated by companies (such as pineapple companies in Ishigaki) and by institutions (such as schools and experiment stations). The total acreage of cultivated land in 1964 by farm households and by commercial companies and institutions was 125,043 acres. This was considerably less than the 148,422 acres cultivated in 1940. However, it should be noted that roughly 16,500 acres of pre-World War II agricultural land were used by the United States military bases. Rent for this military-base land was paid to the former owners and some of this land, roughly 4,000 acres, was in turn leased back to Okinawan farmers for cultivation.

The acreage and the ownership of the land cultivated by farm households in various categories in the major areas of the Ryukyu Islands in 1964 are shown on Table 6. This table is equivalent to Table 2 which gave data for 1940. Though most of the cultivated land was in Okinawa, the acreage on Okinawa had decreased very considerably, going from 103,802 acres in 1940 to 74,576 acres in 1964. Cultivated land in Miyako had not decreased greatly, declining from 30,484 acres in 1940 to 26,764 acres in 1964. On the other hand, acreage of cultivated land had increased considerably in Yaeyama. Eighty-six per cent of the cultivated land was in dry (or common) fields, used mostly for sugarcane and to a lesser extent for sweet potatoes. Fourteen per cent of the cultivated land was in rice (or paddy) fields and in a newly introduced crop, pine-

TABLE 6

ACREAGE AND OWNERSHIP OF LAND CULTIVATED BY FARM HOUSEHOLDS, RYUKYU ISLANDS, 1964

	Owned		Leased		Total	
	Dry Fields	Other Fields	Dry Fields	Other Fields	Dry Fields	Other Fields
Northern Okinawa	13,416	5,200	4,802	1,904	18,218	7,104
Central Okinawa	10,683	695	5,880	372	16,563	1,067
Southern Okinawa	24,154	1,630	5,318	522	29,472	2,152
Total Okinawa	48,253	7,525	16,000	2,798	64,253	10,323
Miyako	23,432	16	3,314	2	26,746	18
Yaeyama	10,136	4,067	3,735	2,880	13,871	6,947
Total Ryukyu Islands	81,821	11,608	23,049	5,680	104,870	17,288

SOURCE: 1964 Agricultural Census data as summarized in *1969 Okinawa Statistical Yearbook,* pp. 109–10. Original data in *ares* converted to acres (1 *are* = 0.0247 acres).

NOTE: Dry fields, or common fields, are devoted largely to sugarcane and sweet potatoes. Other fields include paddy fields (or rice fields) and fields used for "tree crops," including pineapples. Leased land includes 1,507 acres in northern Okinawa, 2,301 acres in central Okinawa, and 204 acres elsewhere leased from the United States Armed Forces.

apples. Regarding ownership of the land, three-quarters of the cultivated land of the Ryukyu Islands was owned by those who tilled it in 1964.

Data on the acreages devoted to the major crops in 1964 are given in Table 7. Sugarcane was by far the dominant crop. In Ishigaki (in Yaeyama) and in northern Okinawa pineapples also were significant. Sweet potatoes, which had once occupied such a relatively large area, were of secondary importance. Chart 2 contrasts the acreages of major crops in 1937 and 1964 for Okinawa, Miyako, and Yaeyama. Study of the chart indicates some of the major shifts in cultivation which took place in this period of time.

According to the 1964 Agricultural Census there were 72,548 farm households which cultivated more than 5 *ares* of land. This was a decided decrease from the figures of 85,235 farm households in 1937, 82,817 farm households in 1951, and 84,353 farm households in 1955, counted on the same basis of cultivating more than 5 *ares*. There were 32.4 farm households per square kilometer of total land area in the Ryukyu Islands in 1964 as contrasted to 38 farm households in 1937 and 37 farm households per square kilometer of total area in 1951.

TABLE 7

ACREAGE AREAS OF MAJOR CROPS, RYUKYU ISLANDS, 1964

	Northern Okinawa	Central Okinawa	Southern Okinawa	Total Okinawa	Miyako	Yaeyama	Total Ryukyu Islands
Rice							
1st Crop	2,707	321	766	3,794		2,388	6,182
2d Crop	2,297	398	862	3,557		309	3,866
Total	5,004	719	1,628	7,351		2,697	10,048
Irish potatoes	354	393	430	1,177	262	184	1,623
Soybeans	23	38	50	111	884	131	1,126
Sweet potatoes	2,701	2,508	1,913	7,122	1,650	674	9,446
Sugarcane	14,610	12,115	20,496	47,221	17,562	8,899	73,682
Pine-apples	3,585	143	334	4,062	6	5,258	9,326
Total	26,277	15,916	24,851	67,044	20,364	17,843	105,251

SOURCES: Various tables drawn from the 1964 Agricultural Census and given in *1968 Okinawa Statistical Yearbook* (Naha: Government of the Ryukyu Islands, 1970). Data in *ares* and converted to acres (1 *are* = 0.0247 acres). Wheat and small grains are negligible in amount, totaling roughly 600 acres for the Ryukyu Islands. Vegetables have some significance in agricultural production, but acreages are small.

The number of farm households per square kilometer of total area for each of the 59 subdivisions of the Ryukyu Islands in 1964 is given on Map E. The pattern has some similarities to that of 1951 as shown on Map D but also some significant differences. The eastern side of northern Okinawa had very few farm households. A large part of this area was used as training grounds by the United States Armed Forces. Since 1951, all of northern Okinawa had a decrease in number of farm households. The offshore islands of Okinawa lost farm populations; some of these losses were relatively great. For example, Tokashiki in the Kerama group had 312 farm

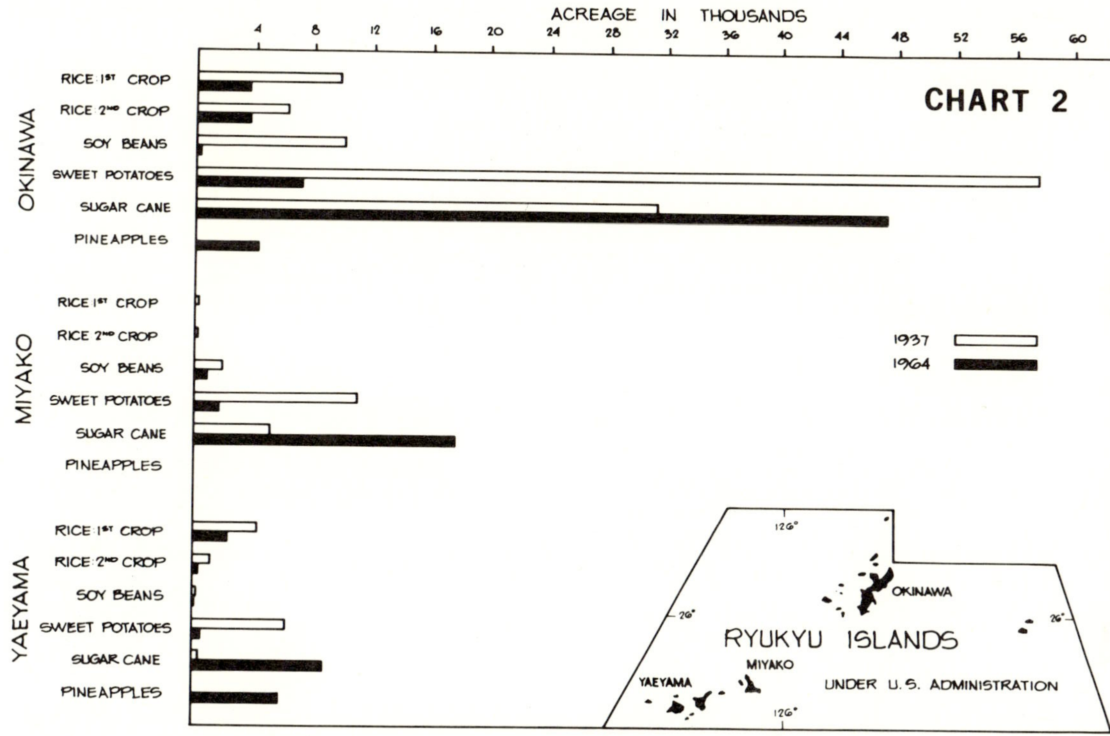

CHANGES IN ACREAGE OF MAJOR CROPS

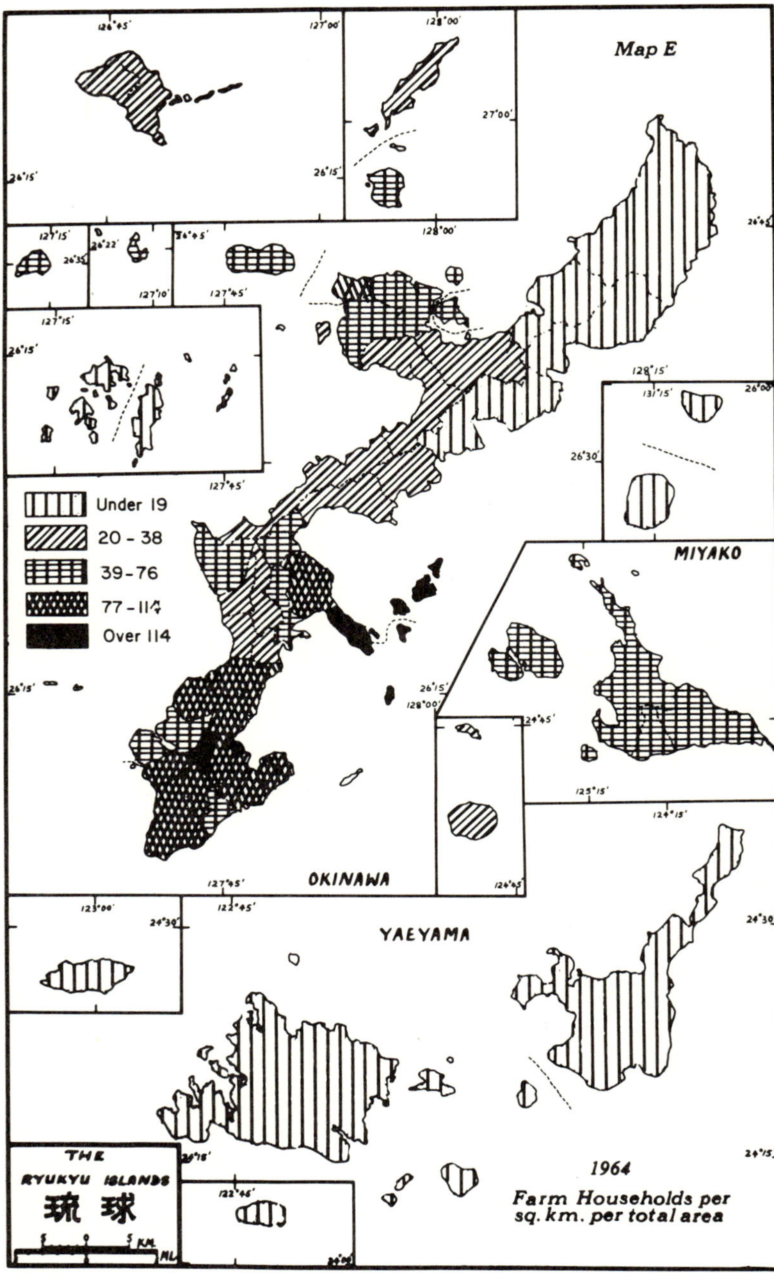
Map E
Under 19
20 - 38
39-76
77-114
Over 114
MIYAKO
OKINAWA
YAEYAMA
THE
RYUKYU ISLANDS
琉球
KM.
1964
Farm Households per
sq. km. per total area

households in 1951 but less than half that number, 152, in 1964. The Daito Islands had very slight gains as did Ishigaki and Yonabaru, east of Naha. In the area around Kadena Air Base and on Ie Island (the site of another U.S. military air field, where farmers were being allowed to lease land for cultivation) the decrease in farm households was not so great. In the cases of Yomitan and Chatan, two political subdivisions near Kadena Air Base, there were actually small increases in numbers of farm households. With these exceptions every political subdivision in the Ryukyu Islands had fewer farm households in 1964 than in 1951.

Paddy rice is a specialized crop in the Ryukyu Islands with production only in certain areas favored by soils and terrain. In some of these areas rice may be double cropped because of the length of the subtropical growing season. As may be noted from Map F, a major area of production in 1964 was on the small alluvial plains in northern Okinawa. The area north of Nago had long had a reputation of being the "rice-bowl" of Okinawa, but conversion of former paddy fields to sugar production and the difficulties of securing sufficient labor for rice cultivation had decreased the importance of rice production in this area. On some of the offshore islands rice fields producing supplies for local consumption occupied a significant percentage of the cultivated land. On the limestone-based soils of southern Okinawa, the Daito Islands, and Miyako, rice was virtually absent.

The distribution of pineapple production in 1964, as noted on Map G, was concentrated in northern Okinawa, on Kume Island, and in Ishigaki, where the soils derived from metamorphic and igneous rocks are suitable for pineapple cultivation. This crop was introduced comparatively recently on a commercial basis in the Ryukyu Islands. It was processed at small canneries, the canned pineapple being shipped to Japan and sold at a highly subsidized price.

Although sugarcane was a crop which had long been established in the Ryukyu Islands, there was a very rapid increase in its cultivated land. Rice and pineapples only occupied significant percentages of the cultivated land where sugarcane could not be grown. Table 8 gives data on the production of centrifugal and non-centrifugal sugar in metric tons from 1950–51 to 1964–65. This table may be considered a continuation of Table 1 which gave data on sugar production from 1918 to 1940. As already noted, sugarcane

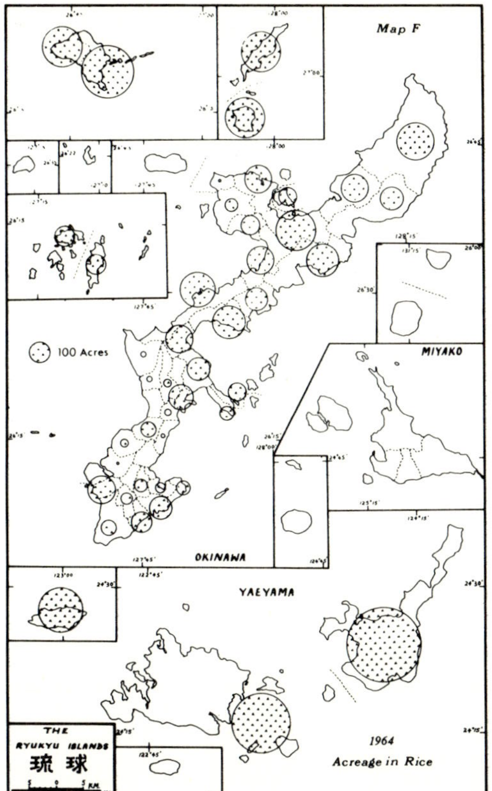
Map F
100 Acres
MIYAKO
OKINAWA
YAEYAMA
THE
RYUKYU ISLANDS
琉球
1964
Acreage in Rice

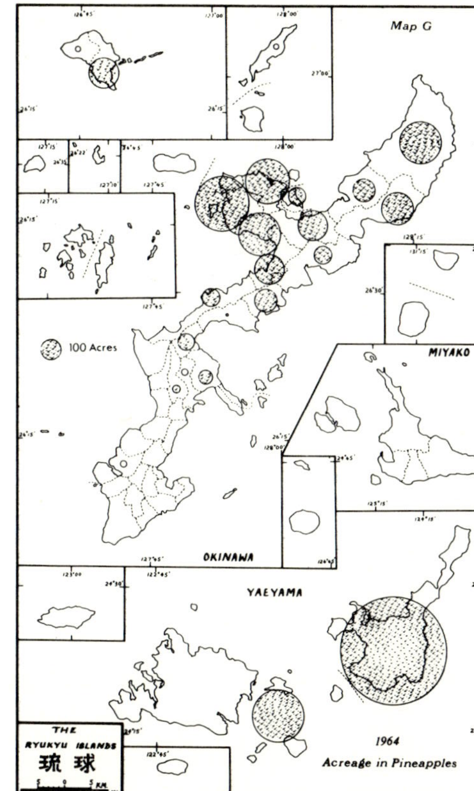
Map G
100 Acres
MIYAKO
OKINAWA
YAEYAMA
THE
RYUKYU ISLANDS
琉球
1964
Acreage in Pineapples

TABLE 8

SUGAR PRODUCTION, RYUKYU ISLANDS, 1950–65
(in metric tons)

Crop Years	Centrifugal Sugar	Noncentrifugal Sugar	Total Sugar
1950–51	0	6,700	6,700
1951–52	1,043	7,430	8,473
1952–53	3,618	14,580	18,198
1953–54	5,639	17,788	23,427
1954–55	10,713	26,917	37,630
1955–56	11,721	39,627	51,348
1956–57	10,105	26,174	36,279
1957–58	11,444	32,774	44,218
1958–59	12,573	43,642	56,215
1959–60	20,795	26,402	47,197
1960–61	57,021	20,125	77,146
1961–62	88,470	23,688	112,158
1962–63	150,494	20,343	170,837
1963–64	128,365	11,121	139,486
1964–65	274,103	24,245	298,348

SOURCES: Tomotaka Ishimine, *Japanese Tariff Preference and the Okinawan Sugar Industry,* Table 4, p. 53. The data were derived from the annual *Statistical Yearbooks,* Agency of Statistics, Planning Department, Government of the Ryukyu Islands.

production was virtually halted in the Ryukyu Islands in the immediate post-war period. Increase in sugarcane production after 1950 was made possible by the high price-subsidy system which was sponsored by the Japanese government.

In addition to the three major crops of rice, pineapples, and sugarcane, there were many other crops grown in the Ryukyu Islands in 1964. Though formerly sweet potato was a major crop, by 1964 production had decreased drastically. However, sweet potatoes were still of some importance in northern Okinawa and in some of the offshore islands. Vegetables occupied a significant percentage of the cultivated land around the urban areas of central and southern Okinawa. In northern Okinawa small areas were devoted to tea production as well as fruit crops such as bananas, which were also grown in the central region, but these did not occupy significant acreages.

6. Part-time Farming and Scale of Farming, 1964

THERE has been a long tradition of part-time farming in the Ryukyu Islands.[23] Many people primarily occupied in small shopkeeping or in local government service carry on agricultural production on a part-time basis. Table 9 gives data on the number of full-time farmers from 1940 through 1965 for the island of Okinawa and its immediate offshore islands. In 1964 the trend toward part-time farming was associated with the increasing availability of employment by the United States Armed Forces, in civil service positions, and in service industries of various kinds. Another significant factor bearing on the increase of part-time farming in the Ryukyu Islands was the heavy emphasis upon the cultivation of sugarcane, which was labor-demanding at only certain times of the year. The decline of sweet potato production and the substitution of imported foods for home consumption were additional factors fostering this trend toward part-time farming.

In 1964 there were 77,129 households in the Ryukyu Islands which were classed as farm households, though some of them cultivated only small amounts of land. Of these, only 23,883 farm households (31 per cent) were classed as making "a specialty of farming." The rest, 53,246 farm households (69 per cent), were engaged in part-time farming. Of this remainder, 19,581 farm households (25 per cent of the total) had farming as a primary—though not sole—occupation; farming was a secondary occupation in 33,665 farm households (44 per cent of the total). Table 10 gives data on the occupations from which the part-time farm households gained

23. The statistics on part-time farming in the Ryukyu Islands follow the definition of the Japanese Ministry of Agriculture and Forestry: "a farm is regarded as a part-time farm if any member belonging to the farm household earns an off-farm income." Part-time farming is an important factor of agriculture in Japan. An interesting account, though with emphasis upon theoretical aspects, particularly "its subjective equilibrium position," is in a chapter by Takeo Misawa, "An Analysis of Part-time Farming in the Post-War Period," in *Agriculture and Economic Growth: Japan's Experience*, ed. Kazushi Ohkawa, Bruce F. Johnston, and Hiromitsu Kaneda (Princeton University Press and University of Tokyo Press, 1970), pp. 250–69. A subsequent paper by Takeo Misawa, Professor of Agricultural Economics of the University of Tokyo, gives more details and relates part-time farming to the "labour allocation in agriculture": "Characteristic Features of Part-time Farming in Japan," *International Journal of Agrarian Affairs* 5 (July 1968): 311–24.

TABLE 9

FULL-TIME FARMERS IN OKINAWA AND ITS ASSOCIATED ISLANDS, 1940–65

Date	Total Number of Farmers	Full-time Farmers	Per Cent of Full-time Farmers	Number As a Per Cent of 1940
1940	251,771	186,850	74.2	100.0
1950	235,147	163,569	57.4	87.5
1955	327,100	171,800	52.5	91.5
1960	350,559	147,481	42.1	78.9
1965	350,484	108,937	31.1	58.0

SOURCE: This table is drawn from page 6 of a student senior thesis in manuscript form at the Department of Geography, The University of the Ryukyus, Shuri, Naha, Okinawa, on the subject *Regional Study of Part-time Farmers on Okinawa Island.*

TABLE 10

OCCUPATIONS OF PART-TIME FARM HOUSEHOLDS, RYUKYU ISLANDS, 1964

Categories	Northern Okinawa	Central Okinawa	Southern Okinawa	Miyako	Yaeyama	Total Ryukyu Islands
Clerical and technical						
General public	1,361	1,042	1,379	555	401	4,738
Civil service	670	796	1,114	258	264	3,102
U.S. Armed Forces	90	609	286	10	3	998
General						
General public	250	157	273	81	77	838
Civil service	1,964	5,406	4,633	643	492	13,138
U.S. Armed Forces	776	5,955	1,596	50	3	8,380
Temporary						
30–100 days	1,934	646	994	1,022	841	5,437
Over 100 days	1,821	732	1,128	1,415	760	5,856
Self-supporting						
Forestry	604	4	2	9	57	676
Fishing	541	252	437	606	194	2,030
Commerce	615	788	725	391	273	2,792
Other	1,241	1,368	1,748	547	357	5,261
Total	11,867	17,755	14,315	5,587	3,722	53,246
Farming specialty	6,600	4,524	6,807	4,070	1,882	23,883
Grand Total	18,467	22,279	21,122	9,657	5,604	77,129

SOURCE: *1969 Okinawa Statistical Yearbook,* table on pp. 81–84. Derived from 1964 Agricultural Census of the Ryukyu Islands.

a livelihood in addition to farming for the major areas of the Ryukyu Islands in 1964.

Though it is sometimes said that part-time farming in Okinawa is associated solely with employment by the United States Armed Forces, it is worth noting that in 1964 only 17.6 per cent of the farm households in the Ryukyu Islands gained part of their livelihood from this source. The percentages of part-time farm households in the political subdivisions near Kadena Air Base were much higher than the percentages for the Ryukyu Islands as a whole; this was to be expected. It is obvious from Table 10 that part-time farming was common among civil servants in 1964. Civil service employment was the supplementary occupation for 30.5 per cent of the farm households.

Map H notes the geographical variation in the percentage of farm households which made a specialty of farming in 1964. The converse of these data would be those who were part-time farm households. Map H notes the relatively high percentage of full-time farmers in northern Okinawa and the Motobu Peninsula, in Miyako, in Kume, in the Daito Islands, and in some of the small offshore islands, such as Aguni. On Aguni, 76 per cent of the 425 farm households were engaged solely in farming in 1964. As might be expected, the relatively small percentage of full-time farm households and conversely the large percentage of part-time farm households were in the urban areas of the Ryukyu Islands, especially around the United States Armed Forces bases in central and southern Okinawa, and in the vicinity of Naha. The city of Naha in the October 1965 census had 257,177 residents, or 27.5 per cent of the total population of the Ryukyu Islands. The urbanization of the Naha-Koza area had a profound effect on the distribution of farm households engaged in part-time farming.

In the Ryukyu Islands the plots of land cultivated by the farm households have traditionally been small and scattered. This pattern enabled the farmer to be protected from total disaster if his fields were damaged by typhoon winds or heavy rainfall associated with typhoons. The small scale of the farms may be noted in Table 11 which gives data on the number of fields cultivated by the 77,129 farm households in 1964 in the major areas of the Ryukyu Islands. It will be noted that two-thirds of the farm households cultivated from one to five plots of land. In the mountainous areas of northern Okinawa the farms were separated into a relatively large number

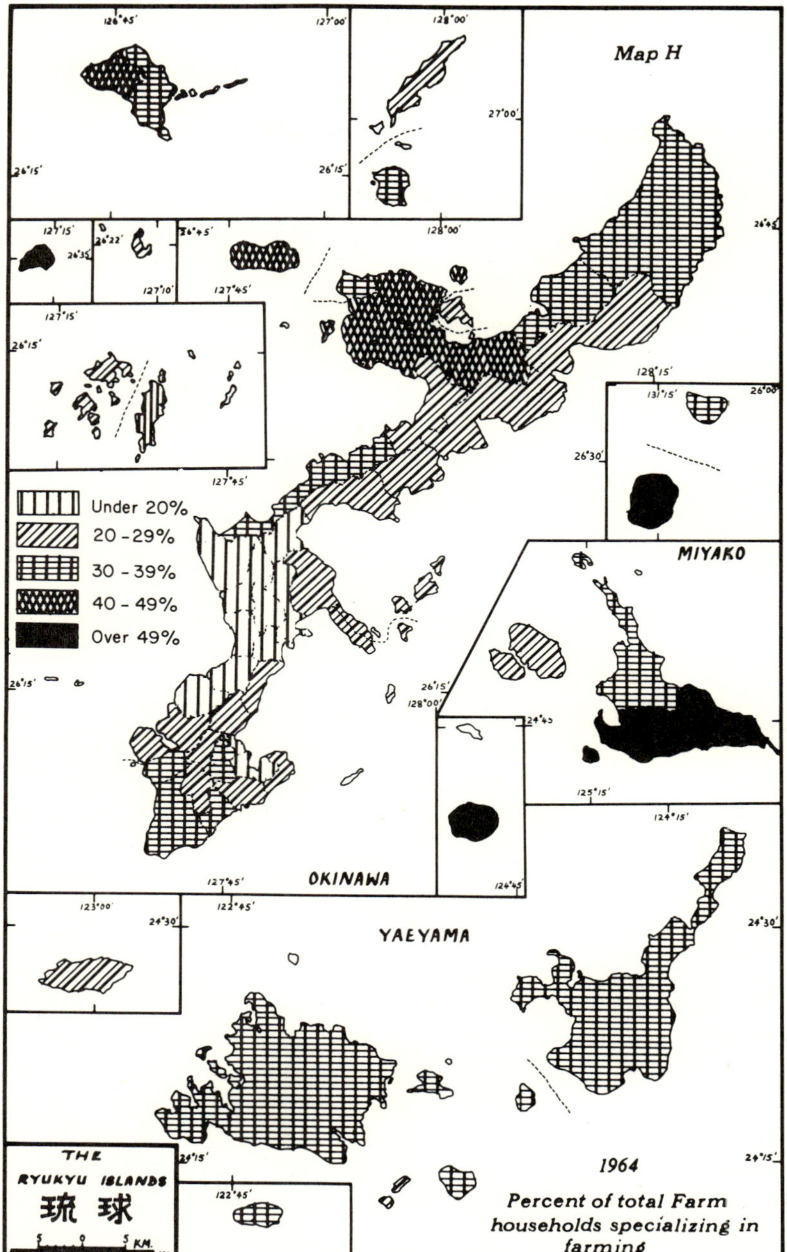
Map H
Under 20%
20-29%
30-39%
40-49%
Over 49%
OKINAWA
MIYAKO
YAEYAMA
THE
RYUKYU ISLANDS
琉球
KM.
ML.
1964
Percent of total Farm households specializing in farming

of plots with 30 per cent of the farm households in that area having from six to ten plots.

The cultivated plots were traditionally scattered in various directions from the village location of the farmstead. According to the *1964 Agricultural Census Report of the Ryukyu Islands,* of the 364,662 individual plots, 43 per cent were less than one-half kilometer from the farmhouse of the farmer who cultivated the plot,

TABLE 11

PLOTS OF LAND CULTIVATED BY FARM HOUSEHOLDS, RYUKYU ISLANDS, 1964

Area	0 Plots	1–5 Plots	6–10 Plots	11–15 Plots	Over 16 Plots	Total
Northern Okinawa	445	9,699	5,555	1,810	958	18,467
Central Okinawa	814	16,571	3,783	700	411	22,279
Southern Okinawa	1,172	13,461	4,969	1,069	451	21,122
Total Okinawa	2,431	39,731	14,307	3,579	1,820	61,868
Miyako	168	7,515	1,776	173	25	9,657
Yaeyama	261	3,823	1,296	188	36	5,604
Total Ryukyu Islands	2,860	51,069	17,379	3,940	1,881	77,129

SOURCE: *1964 Agricultural Census Report, General Summary* (Naha: Government of the Ryukyu Islands, 1966), vol. 2, table on p. 33.

35.3 per cent were between one-half and one kilometer distant, 15.8 per cent were between one and two kilometers away, and 5.9 per cent were over two kilometers distant. Table 12 gives the distance data for the major areas of the Ryukyu Islands in 1964. Distances from the farmhouse were especially great in the southern islands, particularly in northern Ishigaki. In Yaeyama as a whole, 6 per cent of the plots were more than eight kilometers from the farmhouse, and 8 per cent were between four and eight kilometers distant.

The scale of farming, which is determined by the size of the plots and holdings of individual farm households, varies greatly within the Ryukyu Islands. Some of the rice fields are only a few square

TABLE 12

DISTANCE FROM FARMHOUSE TO CULTIVATED PLOTS, RYUKYU ISLANDS, 1964

	Northern Okinawa	Central Okinawa	Southern Okinawa	Miyako	Yaeyama	Total Ryukyu Islands
Less than 0.5 km	52,055	39,109	45,251	14,075	6,205	156,695
0.5–1.0 km	40,521	32,227	38,035	11,955	6,131	128,869
1–2 km	17,352	15,316	12,531	7,860	4,534	57,593
2–4 km	4,958	3,779	2,022	2,980	2,685	16,424
4–8 km	775	351	146	306	1,797	3,375
Over 8 km	92	127	122	21	1,344	1,706
Total	115,753	90,909	98,107	37,197	22,696	364,662

SOURCE: *1964 Agricultural Census Report,* table on 2:34.

feet in size. In contrast, newly developed fields for sugarcane and pineapple in the southern Ryukyus are from two to three hundred yards in length and width and well-suited to mechanized agriculture. However, in 1964 the average size of plots for all of the Ryukyu Islands was only 13.6 *ares* or about one-third of an acre. For more efficient utilization, government-sponsored programs were urging the consolidation of plots. Table 13 gives the data on the scale of farming in the major regions of the Ryukyu Islands in 1964.

TABLE 13

SCALE OF AGRICULTURE, RYUKYU ISLANDS, 1964

Area	Total Cultivated Land (in *ares*)	Number of Plots	Average Size of Plots (in *ares*)
Northern Okinawa	1,025,506	115,753	8.9
Central Okinawa	713,799	90,909	7.9
Southern Okinawa	1,280,308	98,107	13.1
Total Okinawa	3,019,613	304,769	9.9
Miyako	1,083,493	37,197	29.1
Yaeyama	842,870	22,696	37.1
Total Ryukyu Islands	4,945,976	364,662	13.6

SOURCE: *1964 Agricultural Census Report,* table on 2:35.

7. Animal Husbandry, 1964

THE agriculture of the Ryukyu Islands has traditionally been closely associated with animal husbandry. Pigs, kept in sties in the farmyard, were common in the Ryukyu farm household. They were fed mainly from the leaves and tubers of sweet potatoes as well as refuse and garbage from the household. Goats, also fed on the cuttings from the sweet potatoes, were allowed to feed in grasslands and in the sugarcane fields which had been cut. Though chickens were kept in the farm household largely for their eggs, their meat was eaten on festive occasions. Other animals such as cattle and horses were used for plowing and heavy work on the farm as well as for meat products. Often the work animals were rented out or exchanged for human labor by more affluent farm households. It should be noted that during the Battle of Okinawa many cattle were killed; some years elapsed before the number of livestock reached pre–World War II figures.

Other animals were quite distinctive in the animal husbandry of the Ryukyu Islands. Some bulls were prized for their skill in bullfighting, a rather specialized type of spectator and betting sport in Okinawa. On Yonaguni Island a breed of small horses was developed. In the Yaeyama Islands some water buffalo were used as work animals in the wet paddy fields.

As commercialization of agriculture increased, various new forms of animal husbandry became more significant in the Ryukyu Islands. Commercial production of chickens and eggs for the United States Armed Forces as well as for the Ryukyuan market gained in economic importance. However, it was not until 1967 that this market expanded greatly. In 1965, approximately 26 thousand eggs were sold commercially, whereas in 1967, approximately 24 million eggs were sold to the United States Armed Forces. Commercial cattle-grazing in northern Okinawa and in some of the offshore islands was started in 1964. An attempt was made to develop feedlots for cattle imported from Australia and destined for the Ryukyuan and Japanese markets, but this project was not successful. With the import of hand tractors and of small trucks, the need for cattle and horses for farm work diminished.

The data in Table 14 indicate the changes in the numbers of livestock between 1939, 1952, 1964, and 1971 for the major areas

TABLE 14

Livestock and Poultry in the Ryukyu Islands, 1939–71

Area	1939	1952	1964	1971
Okinawa				
Dairy cattle	322	88	961	2,221
Beef cattle	21,526	6,749	7,181	10,937
Horses	N.A.	5,110	5,527	2,255
Swine	113,450	107,077	129,612	173,724
Goats	105,521	58,472	37,872	21,179
Chickens	190,636	133,246	1,071,848	2,050,598
Miyako				
Dairy cattle	5		36	49
Beef cattle	4,543	466	5,941	7,783
Horses	N.A.	8,534	7,079	3,292
Swine	13,075	6,580	11,342	7,083
Goats	9,652	20,126	13,389	6,103
Chickens	33,304	37,115	28,790	81,361
Yaeyama				
Dairy cattle	9	1	35	83
Beef cattle	4,084	2,581	6,365	7,549
Horses	N.A.	2,740	1,790	590
Swine	7,453	6,096	5,660	6,321
Goats	2,875	6,306	5,325	2,237
Chickens	19,704	20,288	32,945	90,919
Total Ryukyu Islands				
Dairy cattle	336	89	1,032	2,353
Beef cattle	30,153	9,796	19,487	26,269
Horses	N.A.	16,384	14,396	6,137
Swine	133,978	119,753	146,614	187,128
Goats	118,048	84,904	56,586	29,519
Chickens	243,644	190,649	1,133,583	2,222,878

Sources: *1968 Okinawa Statistical Yearbook*, Table 82, p. 181; *1971 Okinawa Statistical Yearbook* (Naha: Okinawa Prefecture, 1973), Table 33, p. 158.

of the Ryukyu Islands. Maps I, J, K, and L show the distribution of dairy and beef cattle, swine, and chickens in the 59 political subdivisions of the Ryukyu Islands in 1964. There were many work animals in Miyako and Ishigaki; dairy cattle, though not yet in large numbers, were centered in the urban areas of central and southern Okinawa. These areas also provided markets for pork so that there were a large number of piggeries in the environs of Naha. A close correlation between the distribution of sweet potatoes and pigs had existed throughout the Ryukyu Islands, but this began

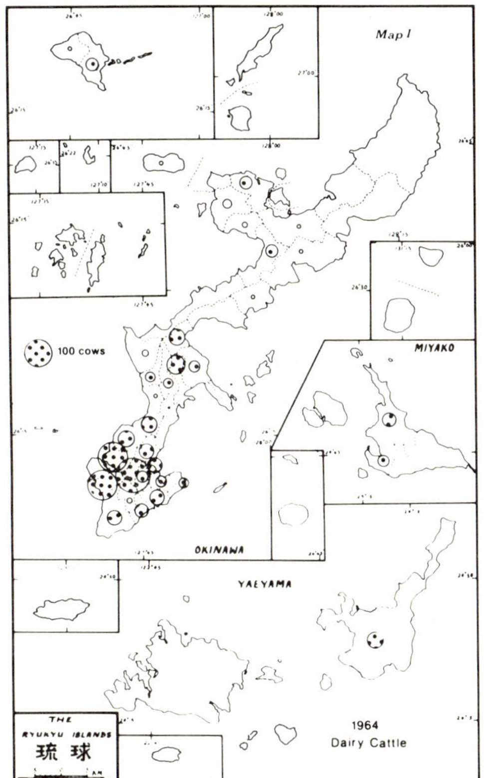
Map I
100 cows
MIYAKO
OKINAWA
YAEYAMA
THE
RYUKYU ISLANDS
琉球
1964
Dairy Cattle

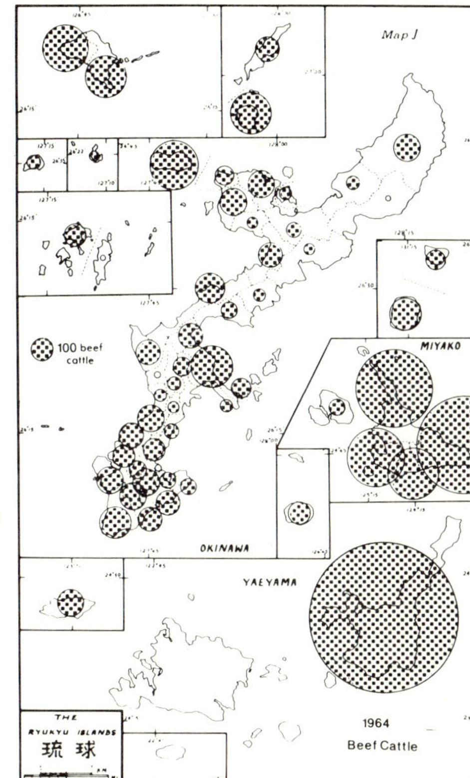
Map J
100 beef cattle
MIYAKO
OKINAWA
YAEYAMA
THE
RYUKYU ISLANDS
琉球
1964
Beef Cattle

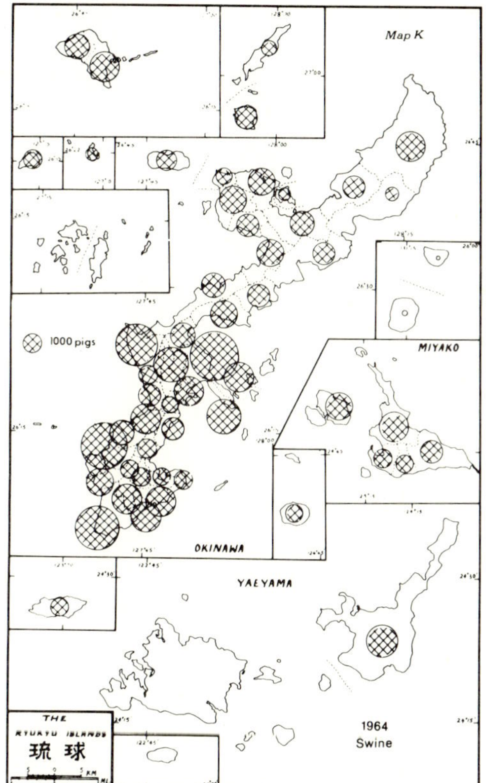
Map K
1000 pigs
MIYAKO
OKINAWA
YAEYAMA
1964
Swine
THE
RYUKYU ISLANDS
琉 球

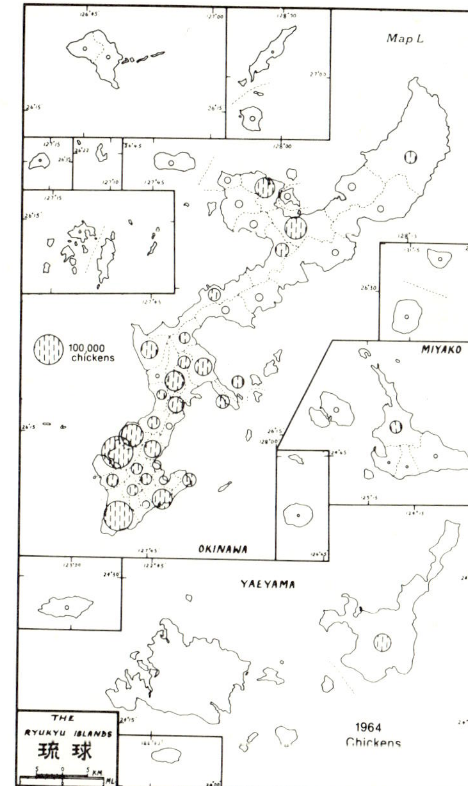
Map L
100,000
chickens
MIYAKO
OKINAWA
YAEYAMA
1964
Chickens
THE
RYUKYU ISLANDS
琉 球

to change with the development of piggeries near urban markets. Chickens were almost ubiquitous in the farm household; the commercial chicken and egg production, however, was concentrated in Haneji-son in northern Okinawa and near Naha and Koza in central Okinawa. Horses were raised mainly in the more remote areas such as the western Motobu Peninsula, Ie, Iheya, Izena, and Aguni islands, and in Miyako and Ishigaki in the southern Ryukyus. Water buffaloes were distributed in appreciable numbers in southeastern Okinawa and in Ishigaki and Yonaguni islands. Goats were widely distributed, with major concentrations in the Motobu Peninsula, the islands offshore from Okinawa, and in Miyako and Ishigaki.

A rural road winds among sugarcane fields and passes near an old tomb in southern Okinawa.

A small hand tractor and a diesel-powered water pump in an area of vegetable and sugarcane cultivation in southern Okinawa.

A water buffalo being used to prepare a paddy field in which the rice shoots growing in a seed bed in the background will be transplanted.

Cattle grazing and cooling off in a small pool in Yonaguni in the southern Ryukyu Islands.

An isolated farmstead in the rolling hills of southern Okinawa where sugarcane is the dominant land use.

The village of Ada in northern Okinawa with terraced sweet potato fields on the hill slope. In the lower right is the agricultural cooperative store. Sugarcane is growing in former paddy fields on the valley floor.

8. Changes in Agriculture between 1964 and 1971

BY 1964, commercialization of agriculture had increased greatly in the Ryukyu Islands. The small-scale, self-sufficient farm household, operated as a basis for subsistence rather than monetary gain, was passing quickly from the scene. On some of the isolated islands, remnants of this type of traditional agriculture still persisted, but even in these places the farming population was decreasing, sometimes drastically, and fields were being abandoned. The attractiveness of off-the-farm employment, the availability of imported foods and the high subsidies for commercial crops, such as sugar and pineapples, were spelling the doom of the "family farm," a trend which accelerated in the decade of the 1960s. It is worthwhile to compare the situation of 1964 with that of 1971, the eve of the reversion of Okinawa to Japan.[24]

Indicative of the major changes that have taken place in the agriculture of the Ryukyu Islands is the decrease in the number of farm households: no matter how small the area which is cultivated,

24. In 1962–64, I served as Civil Administrator of the Ryukyu Islands. During this period I had many opportunities to visit many of the islands and to observe the agriculture of the Ryukyu Islands. Some of the most difficult problems with which the Civil Administration had to deal were in the field of development of the agricultural and human resources of the Ryukyu Islands. In 1967, while a Research Professor of Geography of the University of Vermont, I visited the Ryukyu Islands for some weeks and carried on some field research in some of the agricultural areas. In the summer of 1970, I carried on further field research on geographical aspects of agricultural changes in the Ryukyu Islands. The major agricultural areas were observed; some type-study areas were studied intensively. Since the field research was completed, statistics from the 1970 Population Census and the 1971 Agricultural Census have become available and have been used to quantify some of the changes that have taken place.

all farms are included in this category. In 1964, there were 77,129 such farm households; in 1971, however, the number declined to 60,346. This is a decrease of 22 per cent. The number of persons in farm households decreased from 435,375 in 1964 to 312,809 in 1971. However, it is significant that there was no comparable decrease in the amount of land under cultivation by the farm households. A total of 122,158 acres was cultivated by the farm households in 1964, and 116,411 acres were cultivated in 1971, a decrease of only 4.5 per cent. The geographical differences in these changes are of interest. In some places, such as Miyako, there was an actual, though slight, increase in the acreage of cultivated land. However, a major decrease in cultivated land took place in the urbanized areas of central and southern Okinawa. As will be pointed out subsequently, some local changes of significance occurred in terms of cultivated land and in crops on the land.

To these figures representing land under cultivation by farm households may be added the small fraction of land under cultivation by companies and institutions. This amount also decreased from 2,885 acres in 1964 to 2,734 acres in 1971. The number of such companies and institutions having cultivated land decreased greatly from 477 in 1964 to 265 in 1971. It should be noted that half of this cultivated land (1,338 acres in 1971) was in Ishigaki and was mainly held by pineapple companies.

The changes in farm households and amounts of land cultivated by farm households, companies, and institutions were a reflection of the changing economic and social character of the Ryukyu Islands. Agriculture continued to be less significant in the total economy. According to the official October censuses, the population as a whole had increased slightly from 934,176 in 1965 to 945,111 in 1970. However, the urbanized areas of southern and central Okinawa, extending northward from Naha to Koza and comprising less than 5 per cent of the total area of the Ryukyu Islands, increased in population from 338,451 in 1965 to 426,654 in 1970. Forty-five per cent of the total population of the Ryukyu Islands lived in this urban area. The total number of employed persons over 15 years of age in the Ryukyu Islands was 350,414 in 1965. Of these, 109,698 (31.3 per cent) were primarily employed in agriculture and forestry. There were 357,440 persons over 15 years of age employed in 1970, and of these, 72,287 (20.2 per cent) were primarily employed in agriculture and forestry. Although the gross national income increased dramatically during this period, the share contributed by agriculture and forestry did not increase greatly. In 1964, the gross national income of the Ryukyu Islands was estimated at $296,900,000; of this, $40,800,000 (13.74 per cent) was contributed by agriculture and forestry. In 1970, the gross national income had increased greatly to $753,000,000; but of this only $57,700,000 (7.66 per cent) was derived from agriculture and forestry.[25]

The basic causes for these changes in the agriculture of the Ryukyu Islands have already been indicated. The sugar price stabilization program of the Japanese government was a major factor. Although the program was slightly altered in 1967, it still provided a very considerable subsidy to the farmers growing sugarcane in the Ryukyu Islands.[26] However, under these stabilized conditions sugar production did not increase as it had in the decade before 1964 but decreased and leveled off at an average annual production of 225,000 metric tons during the period from 1964 to 1971. It should be noted that the production decreased greatly in 1971–72, in part because of climatic conditions. The 1971–72 production was only 140,843 metric tons. Pineapple production also was highly subsidized by the Japanese government with a resultant considerable increase in acreage devoted to pineapples. The United States Armed Forces stationed in Okinawa continued to contribute greatly to the economy of the Ryukyu Islands. They provided an increasingly important cash market for local vegetables, beef, milk, chickens, and eggs. In addition, the United States government through its programs of food grain exports under Public Law 480,

25. The data for this analysis are derived from various official documents. Among the most significant of these are the annual *Statistical Yearbooks* of the Agency of Statistics, Planning Department, Government of the Ryukyu Islands, Naha; the *Population Census Reports* issued every five years by the Government of the Ryukyu Islands; and the annual *Reports* of the United States Civil Administration of the Ryukyu Islands. The agricultural censuses taken in 1964 and in 1971 provide a wealth of data; though they are in Japanese, the tables may be used without much difficulty. The University of the Ryukyus has published much valuable material and analysis on agricultural and land utilization of the Ryukyu Islands. Particularly useful is a bilingual study prepared in a cooperative effort by faculty of the University of the Ryukyus and of Michigan State University: *Soils and Land Use in the Ryukyu Islands*, by T. Chinzei, K. Oya, and Z. Koza, in collaboration with Roy L. Donahue and John C. Shickluna (Naha, Okinawa: College of Agriculture, University of the Ryukyus, 1967).

26. For these changes see Tomotaka Ishimine, *Japanese Tariff Preference and the Okinawan Sugar Industry.*

Title II, supported school lunch programs and relief food programs for the needy, particularly in times of natural disasters, such as typhoons and droughts.[27]

One of the most profound impacts on the agriculture of the Ryukyu Islands was the development of alternate, and more attractive, job opportunities for the Ryukyuan farmers and particularly the younger members of their families. Universal educational and training programs had helped develop the skills and knowledge of these young people. They were anxious to leave the land and move to the growing urban centers of both Okinawa and Japan. In addition to outside occupations, these young people might assist their parents and older relatives in the farm work on a part-time basis and would, in many cases, still be counted as part of the farm household. Table 15 gives data on the changes in the age groups of the farm household members between 1964 and 1971, according to the agricultural censuses of those years.

The effect of population control measures may be noted in the lessened number of children, a drop from 190,568 children aged 14 and under in 1964, to 104,619 in this age group in 1971. The crest of the post-World War II birth rate of the Ryukyu Islands had moved into the 15–19 year-old age group. The small differences in numbers of older persons, over the age of 60 between 1964 and 1971, despite the considerable decrease in number of persons in farm households, is also significant. Along with the decrease in the

27. The food relief and school lunch programs were administered for many years by the Ryukyu Islands Voluntary Agencies Committee (RIVAC) composed mainly of Christian missionary groups. A brief but interesting summary account of the RIVAC programs is given in *Civil Administration of the Ryukyu Islands, Report for the Period 1 July 1969 to 30 June 1970* (Naha: USCAR, 1971), pp. 205–8. After 1970 the Christian missionary groups, having given devoted service for many years and anticipating the reversion of the Ryukyu Islands to Japan, dissolved RIVAC, and new administrative organizations were formed with Okinawan leadership to take over these responsibilities. According to the USCAR *Annual Report,* the new programs "will provide supplementary food assistance for approximately 246,000 school children," as well as "food commodities for an estimated 8,000 other children in nurseries and institutions." Moreover, these programs provided "relief supplies for 10,291 persons affected by typhoon and flood" in 1970 in the southern islands of the Ryukyus which had been badly damaged by typhoons. Though these relief activities were not always publicly acknowledged, it may be noted that by 1971 almost one-third of the total population of the Ryukyu Islands received some of their daily food from United States assistance programs. The dietary changes occasioned by the school lunch program over the recent years have been profound; no longer are sweet potatoes a staple (or desired) item in the lunch of children and workers; they have been replaced by "sandwichu"!

TABLE 15

AGE OF MEMBERS OF FARM HOUSEHOLDS, RYUKYU ISLANDS, 1964, 1971

Ages	1964 Male	1964 Female	1964 Total	1971 Male	1971 Female	1971 Total
0–14	96,918	93,650	190,568	53,048	51,571	104,619
15–19	18,507	16,870	35,377	20,112	19,035	39,147
20–24	13,012	11,269	24,281	11,078	9,521	20,599
25–29	12,585	11,118	23,703	5,735	4,688	10,423
30–39	21,260	23,293	44,553	15,994	15,366	31,360
40–49	17,747	22,541	40,288	16,102	18,623	34,725
50–59	14,817	18,192	33,009	13,641	16,535	30,176
60–64	5,789	7,551	13,340	5,826	6,905	12,731
Over 64	12,900	17,356	30,256	12,300	16,729	29,029
Total	213,535	221,840	435,375	153,836	158,973	312,809

SOURCES: 1964 data derived from the 1964 Agricultural Census, *1968 Okinawa Statistical Yearbook,* Table 50, p. 136; 1971 data derived from Table 7 in "1971 Agricultural Census Report," Preliminary Summary, mimeographed (Naha, April 20, 1972).

number of farm households, there was a decided decrease in the number of members per farm household. In 1964 the average farm household had 5.64 members; in 1971 it had 5.18 members. Table 16 gives data on the changes in actual numbers of members of farm households between 1964 and 1971.

TABLE 16

NUMBER OF MEMBERS OF FARM HOUSEHOLDS, RYUKYU ISLANDS, 1964, 1971

Persons per Farm Household	1964 Households Number	1964 Households Per Cent	1971 Households Number	1971 Households Per Cent
1	3,382	4.4	3,140	5.2
2	7,828	10.1	7,688	12.7
3	8,061	10.6	6,579	10.9
4	8,237	10.7	7,048	11.7
5	9,216	11.9	8,536	14.2
6	10,281	13.3	8,936	14.8
7	10,527	13.6	7,579	12.6
8	8,519	11.0	5,196	8.6
9	5,545	7.2	2,971	4.9
10	3,058	4.0	1,507	2.5
11+	2,475	3.2	1,166	1.9
Total	77,129	100.0	60,346	100.0

SOURCES: 1964 data from *1964 Agricultural Census Report,* Table 17, 2:123; 1971 data from "1971 Agricultural Census Report," Table 6.

The effect of all of these varied forces profoundly changed the character, and, to a considerable extent, the geographical distribution of agriculture of the Ryukyu Islands on the eve of reversion to Japan. It is of value to summarize some of these changes, particularly in their geographical aspects. It needs to be noted in passing that drought, typhoons, and flood conditions occurring in the southern islands of the Ryukyus in the spring and summer of 1971 had some local influence on the data for 1971, but the general trends are still evident.

The farm households of the Ryukyu Islands use land not only for cultivation but for other purposes, such as wood lots, grazing lands, house lots, and grave sites. The amount of land used for purposes other than cultivation by farm households decreased significantly in 1971 to 33,849 acres. In that year, there were 17,685 acres in forest land (including wood lots); 11,823 acres in grass land which had never been cultivated; and 4,341 acres in formerly cultivated land which had reverted to grass land. One factor in this decrease was the decline in the number of farm work animals, thus decreasing the need for grazing land. Increased use of propane gas instead of wood for household cooking was another factor in this decrease in the amount of land used for purposes other than cultivation.

Chart 3 shows—by means of line graphs for the major regions—the decrease in the number of farm households in the Ryukyu Islands from 1937 to 1971. For the Ryukyu Islands as a whole, the number of farm households per square kilometer of total area decreased from 38.0 in 1937 to 37.0 in 1951, to 32.4 in 1964, and continued to drop drastically to 25.9 farm households per square kilometer of total area in 1971. Map M indicates the number of farm households farming more than 5 *ares* per square kilometer in each political subdivision in 1971. The patterns are generally similar to the maps which have been drawn on the same basis for selected previous years (see Maps C, D, and E). The urban growth of Naha took out of production much of the small area which had been cultivated in that region. The urban and military areas around Kadena Air Base in central Okinawa also continued to have relatively few farm households per square kilometer of total area. In comparing this map for 1971 with the similar map for 1964 (Map E), it is apparent that every political subdivision had fewer farm households per square kilometer of total area in 1971. Map N gives

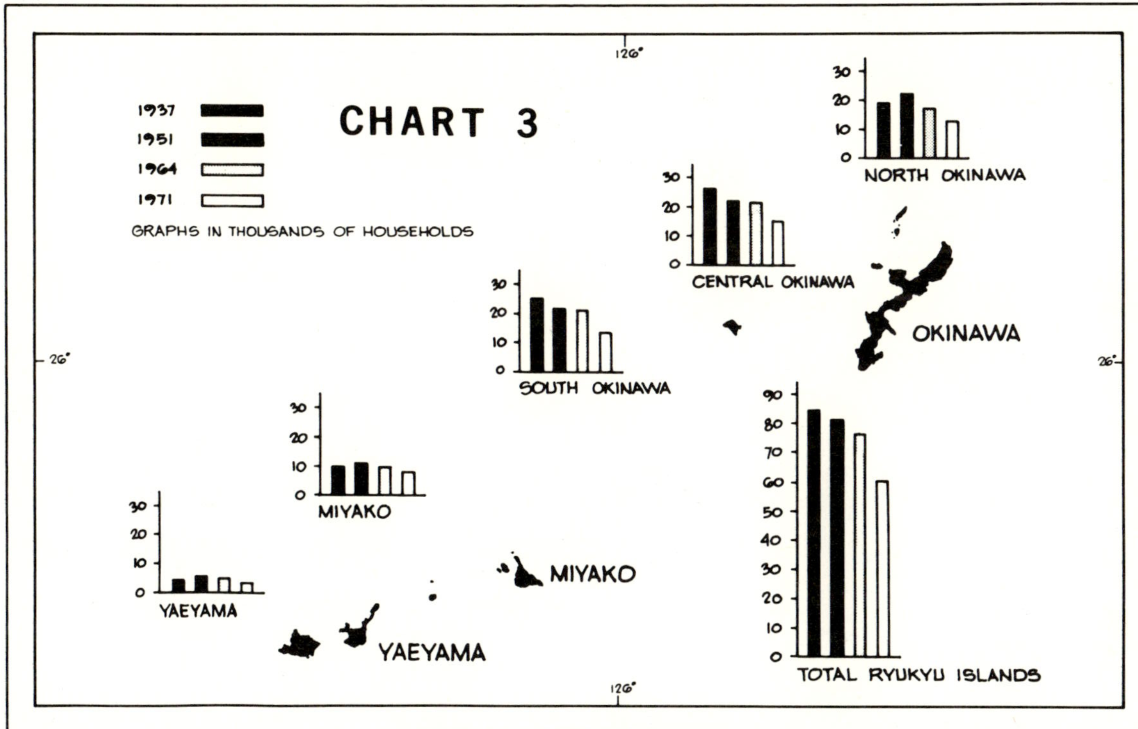

FARM HOUSEHOLDS

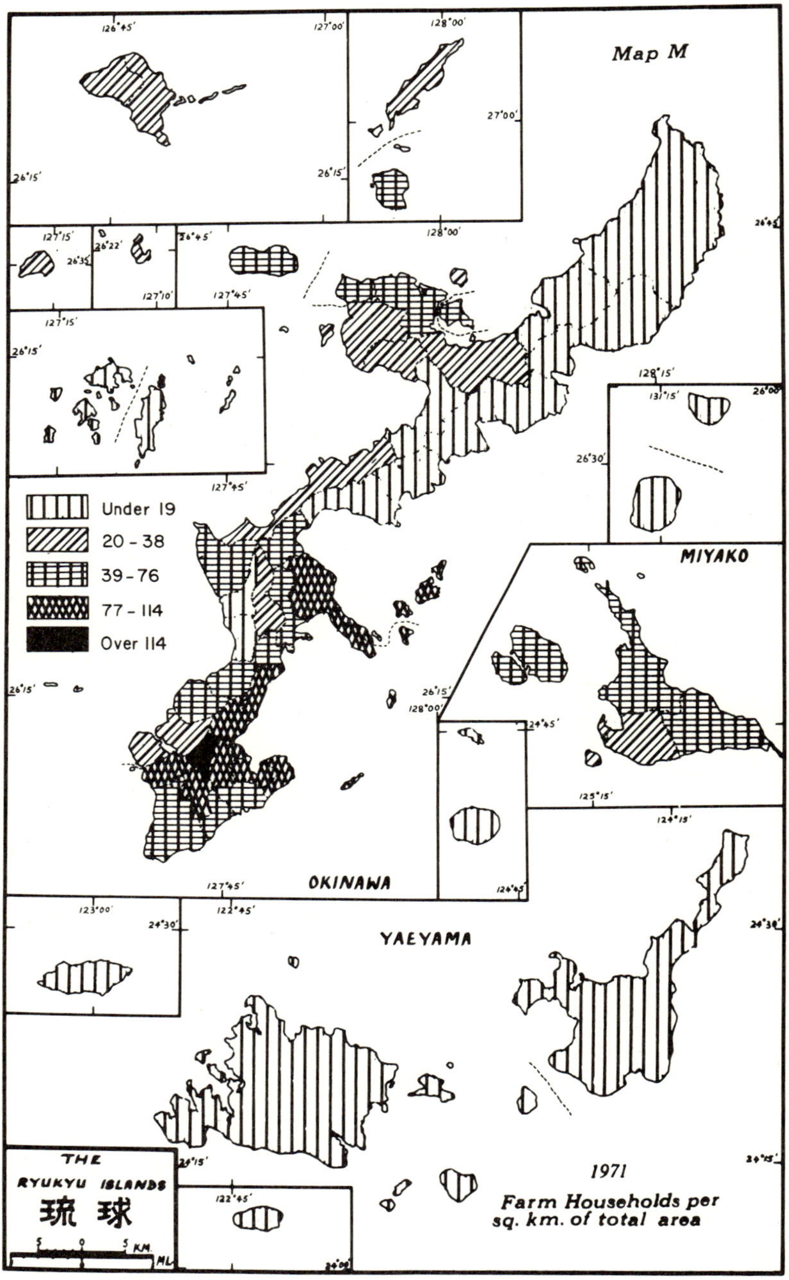
Map M
Under 19
20 - 38
39 - 76
77 - 114
Over 114
OKINAWA
MIYAKO
YAEYAMA
THE
RYUKYU ISLANDS
琉球
KM.
1971
Farm Households per
sq. km. of total area

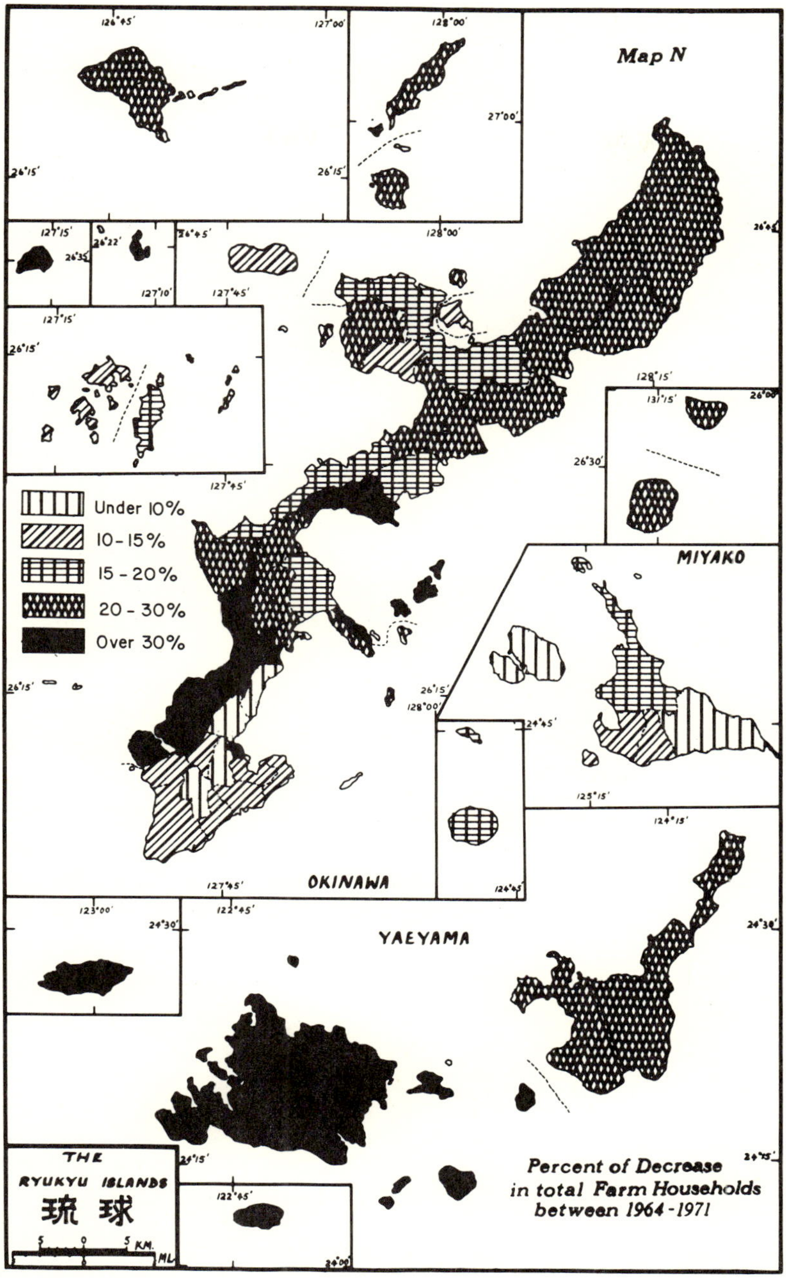
Map N
Under 10%
10-15%
15-20%
20-30%
Over 30%
MIYAKO
OKINAWA
YAEYAMA
THE
RYUKYU ISLANDS
琉球
KM.
ML.
Percent of Decrease
in total Farm Households
between 1964-1971

the percentage of decrease in total farm households between 1964 and 1971 in each of the political subdivisions. The loss was particularly marked in the urbanized area from Naha northward to beyond Koza. There were also marked declines in the islands offshore from Okinawa and in Yaeyama. This trend was to be expected with the increased commercialization of agriculture and the decline in the numbers of "family farms."

Although there was a sharp decrease in the number of farm households in the Ryukyu Islands between 1964 and 1971, a proportionate decrease in the amount of cultivated land in this period did not take place. In fact, where pineapple cultivation had been introduced in the Motobu Peninsula and in northern Okinawa there was an actual increase in land under cultivation. Kushi-son, which had had a relatively small percentage of its land under cultivation in 1964, underwent such a "pineapple boom" that it had a 66.3 per cent increase in cultivated land between 1964 and 1971. Other areas in northern Okinawa and in the Motobu Peninsula where pineapple was being introduced had similar increases, sometimes at the risk of causing serious land erosion. On Ie Island off the Motobu Peninsula, a considerable amount of land was released for cultivation by the United States Armed Forces. In the area east of Naha the growing of fresh vegetables for the urban market gave new uses for cultivated land. Thus, two political subdivisions in that area had net increases in cultivated land. In Miyako and the Daito Islands, sugarcane cultivation was increased; so that these areas, also, had some increase in acreages of cultivated land. Map O shows the percentage of increase or of decrease in land cultivated by farm households between 1964 and 1971. For the Ryukyu Islands as a whole, there was a decrease of 4.5 per cent in cultivated land.

9. Agricultural Situation, 1971

THE agricultural changes in the Ryukyu Islands between 1964 and 1971 resulted, as would be expected, in a pattern of agriculture in 1971 which was quite different from that prevailing in the previous decades. Agriculture was of much less significance in the total economy. Most of the agricultural production moved through

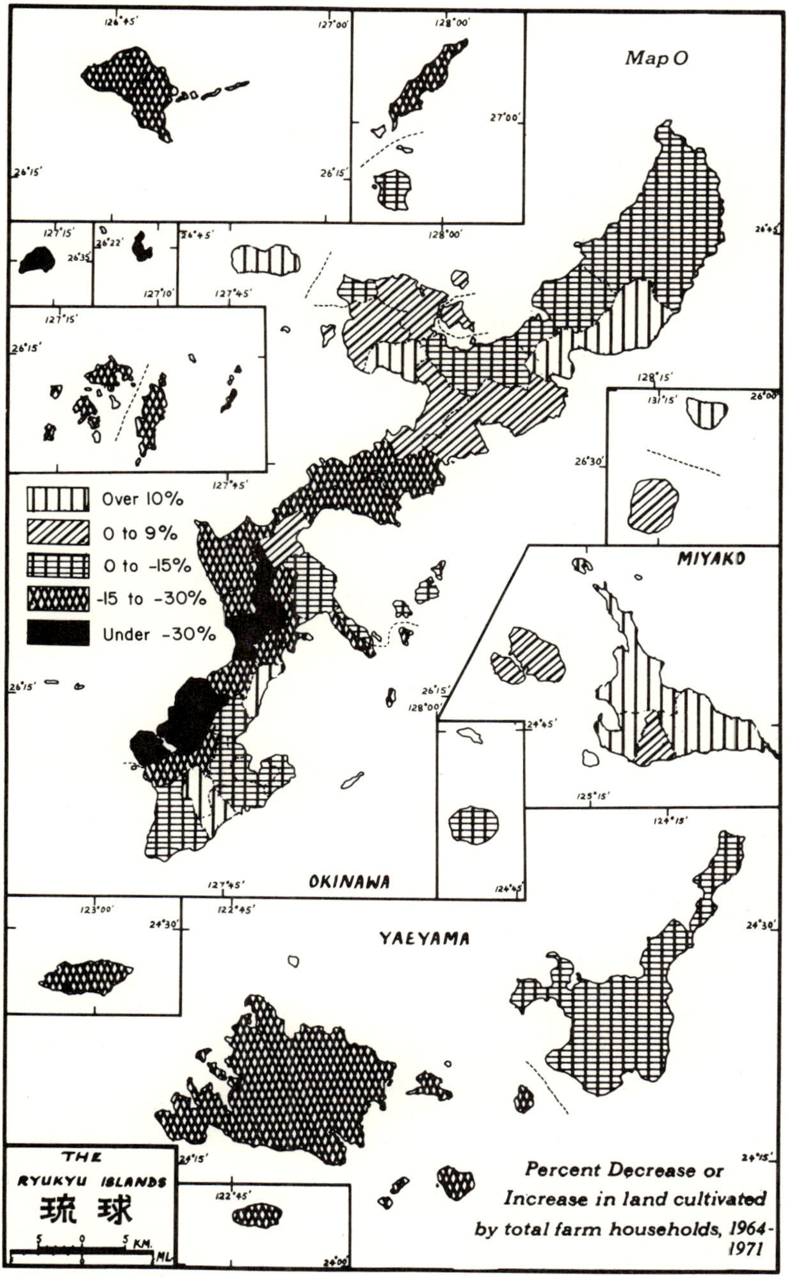

Map O
Over 10%
0 to 9%
0 to -15%
-15 to -30%
Under -30%
OKINAWA
MIYAKO
YAEYAMA
THE RYUKYU ISLANDS
琉 球
KM.
ML.
Percent Decrease or Increase in land cultivated by total farm households, 1964-1971

commercial channels; the self-sufficient "family farm" was found only on some of the isolated islands. Sugar and pineapples destined for price-subsidized markets in Japan dominated the commercial agriculture of the Ryukyu Islands.

The same patterns of ownership of the cultivated land continued, however. Three-quarters of the land was owned by those who tilled the land, mostly small landholders. Table 17 summarizes the data on the nature and ownership of the land cultivated by the 58,683 farm households that were actually cultivating land in 1971.

TABLE 17
NATURE AND OWNERSHIP OF LAND CULTIVATED BY FARM HOUSEHOLDS, RYUKYU ISLANDS, 1971
(in acres)

	Owned			Leased			
Area	Dry Fields	Paddy Fields	Tree Crops	Dry Fields	Paddy Fields	Tree Crops	Total
Northern Okinawa	11,505	1,614	5,318	4,181	573	1,959	25,150
Central Okinawa	9,218	256	120	5,041	61	320	15,016
Southern Okinawa	21,706	763	97	5,219	173	284	28,242
Total Okinawa	42,429	2,633	5,535	14,441	807	2,563	68,408
Miyako	24,437		2	4,692			29,131
Yaeyama	8,187	1,767	3,518	3,004	942	1,506	18,924
Total Ryukyus	75,053	4,400	9,055	22,137	1,749	4,069	116,463

SOURCE: *1971 Okinawa Statistical Yearbook.*

This table may be compared with Table 6 which summarizes the data for 1964. The amount of land leased from the United States Armed Forces had decreased sharply from the 4,012 acres so leased in 1964 to 2,481 acres in 1971, or only 2 per cent of the total cultivated land. This leased land was mainly in two localities: on Ie Island, off the Motobu Peninsula, and in the area surrounding Kadena Air Base in central Okinawa.

In many areas of the Ryukyu Islands, particularly in Okinawa, paddy fields had been converted to sugarcane. The remaining rice fields were concentrated in northern Okinawa and on isolated islands, such as Iheya, Izena, and Kume, offshore from Okinawa, and in Ishigaki, Iriomote, and Yonaguni in the southern Ryukyus. Because these isolated areas had added costs in importing rice it was economically advantageous for the farmers to retain some of their rice crop for local consumption. However, in the more easily accessible areas rice was not so profitable economically, although the Government of the Ryukyu Islands had attempted to bolster rice production through a rice purchase program.

There is a significant difference in the types of cultivated land in the Ryukyu Islands. The preliminary agricultural census for 1971 provided data by three types of fields: dry fields, paddy fields, and fields used for "tree crops"; this last category includes land used for pineapples, and to a very much lesser extent, tea trees and fruit trees, such as bananas and papayas. These data are available in terms of the total cultivated land, including the land cultivated by companies and institutions. Two maps, P and Q, prepared from this data, show the acreage for the paddy fields and for the "tree crops" in each of the subdivisions of the Ryukyu Islands. These maps may be compared with Maps F and G which gave acreage for rice and pineapples in 1964. In 1971, almost one-quarter of all of the rice in the Ryukyu Islands was grown in Ishigaki Island. Small alluvial plains in northern Okinawa and in some of the offshore islands, such as Kume, Iheya, Izena, and Yonaguni, were devoted to paddy fields also. There were small areas of rice fields on alluvial land along the coast in southeastern Okinawa. Pineapple production was concentrated in northern Okinawa, the Motobu Peninsula, on Kume Island, and on Ishigaki.

Dry fields which were mainly devoted to sugarcane cultivation dominated the use of the cultivated land in 1971. Rice and pineapple production in northern Okinawa and in Yaeyama caused some diversification of land use in those regions. Even in those areas, however, sugarcane was the most important crop. On the limestone-based soils of Miyako, sugarcane was particularly dominant; Miyako, though only one-tenth of the total land area of the Ryukyu Islands, had almost one-fourth of the total area of sugarcane cultivation.

After the peak year of production in 1964–65 when 298,348 metric tons of sugar were produced, sugar production declined and leveled off at around 225,000 metric tons for the Ryukyu Islands, during the period from 1964 to 1971. This stabilization of sugar production was of considerable significance in the agricultural situa-

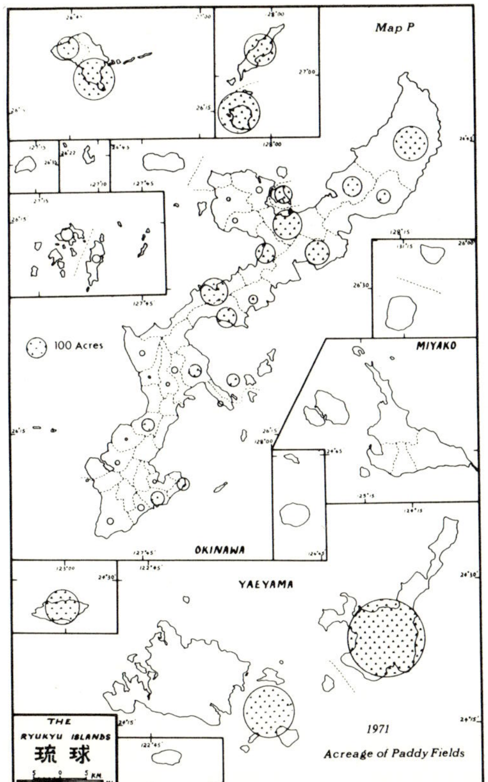

Map P
100 Acres
MIYAKO
OKINAWA
YAEYAMA
THE
RYUKYU ISLANDS
琉球
1971
Acreage of Paddy Fields

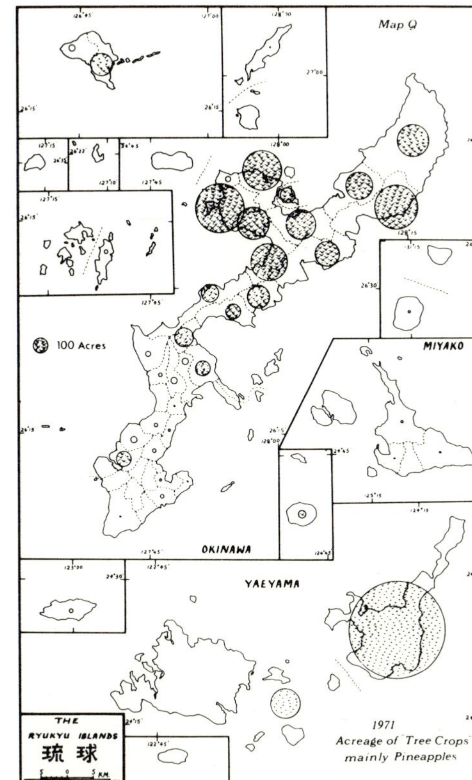

Map Q
100 Acres
MIYAKO
OKINAWA
YAEYAMA
THE
RYUKYU ISLANDS
琉球
1971
Acreage of 'Tree Crops'
mainly Pineapples

tion of the Ryukyu Islands. However, in 1971–72 there was a sharp decline in production; 1971–72 production was only 140,843 metric tons of sugar. The Japanese subsidies had not increased, since world sugar prices had leveled off or declined. Because of shortages of sufficient farm labor, many sugarcane crops were ratooned rather than replanted with more productive new cane. Moreover, climatic conditions, particularly drought and typhoon damage in 1971, were

TABLE 18
SUGAR PRODUCTION, RYUKYU ISLANDS, 1964–72
(in metric tons)

Crop Year	Centrifugal Sugar	Noncentrifugal Sugar	Total Sugar
1964–65	274,103	24,245	298,348
1965–66	210,454	11,389	221,843
1966–67	194,888	10,463	205,351
1967–68	220,667	12,122	232,789
1968–69	212,313	13,852	226,165
1969–70	208,748	10,486	219,234
1970–71	216,173	13,136	229,309
1971–72	136,405	4,438	140,843

SOURCE: *1971 Okinawa Statistical Yearbook.*

unfavorable to increased production. Table 18 gives data on sugar production from 1964 to 1972 and is a continuation of Tables 1 and 8 which give data on production in previous time spans.

10. Part-Time Farming and Scale of Farming, 1971

IN the period from 1964 to 1971 the trend toward part-time farming in the Ryukyu Islands continued to accelerate. A large number of farmers left farming completely. The total number of households classed as farming households, even though some of them may have not been cultivating much land in 1971, was 60,346 households. This was a decided decrease from the equivalent number (77,129) of farm households in 1964. It is worthwhile noting that this decrease was mainly in the number of households which made a specialty of farming and for those part-time households classed as chiefly engaged in farming. Table 19 provides data on this relationship. There was actually a very slight gain in the number of part-time farm households which had only a minor source of income from farming. Put another way, there were probably many part-time farm households which stopped farming altogether, and their places were taken in the statistics by farm households which in 1964 had been chiefly engaged in farming or which had derived the major part of their income from farming. The percentage decrease in full-time farm households, plus those part-time farm households which were chiefly engaged in farming between 1964 and 1971 in the

TABLE 19
CLASSIFICATION OF FARM HOUSEHOLDS, RYUKYU ISLANDS, 1964, 1971

Classification	1964	1971	Change
Specialty in farming	23,883	13,478	−10,405
Part-time, chiefly in farming	19,581	12,253	− 7,328
Part-time, secondarily in farming	33,665	34,615	+ 950
Total	77,129	60,346	−16,783

SOURCES: 1964 and 1971 Agricultural Censuses.

Ryukyu Islands, was 40.8 per cent, a very decided decrease. Variations occurred within the Ryukyu Islands, however; Miyako and the Daito Islands did not have such a decided decrease, but in contrast some of the small islands off Okinawa had very large decreases.

The converse of full-time farming—though it may not be exactly such because the situation is complex—is the increase or decrease in the number of part-time farm households which derive a minor source of their income from farming. As already noted, for the Ryukyu Islands as a whole, the number of such farm households was almost the same in 1964 and 1971. In percentage of total farm households, however, there was a significant increase of those whose income was derived only secondarily from farming. In 1964 there were 33,665 such farm households (43 per cent of the total); in 1971 the number was 34,615 (57 per cent of the total). The major decreases in the numbers of part-time farm households which derived a minor part of their income from farming between 1964 and 1971 took place in the urbanized area from Naha northward to Koza. Other urbanized areas, such as Yonabaru, east of Naha, Nago in northern Okinawa, Hirara on Miyako, and Ishigaki in Yaeyama, also experienced decreases. Decided decreases took place in the

Daito Islands and Yonaguni. In contrast, increases in the percentage of part-time farm households which derived a minor part of their income from farming between 1964 and 1971 took place in the Motobu Peninsula, the outlying areas of Miyako, and in the area south and east of Naha where intensive market gardening was being expanded.

Certain economic advantages are derived from part-time farming in the Ryukyu Islands. There are certain tax advantages for farm households. Membership in the farmers' cooperatives enables persons to buy some commodities at a discount.[28] Cultivation of sugarcane, because it requires labor at only certain times of the year, makes farming as a part-time occupation possible. A significant number of persons with previous experience in farming kept small amounts of land in cultivation, planting vegetables or sweet potatoes in season. Such small-scale farming might be called "nostalgic farming." Other land, though classed as farmland, is left idle in hopes of a speculative profit in land values. Of the total land classed as cultivated in 1971, 4.6 per cent was idle at the time of the agricultural census.

There are some interesting regional differences in the percentages of farmers who made a specialty of farming. Using the 1971 data for 60,346 households classed as farm households—including those which cultivated less than 5 *ares*—the percentage of farm households which made a specialty of farming in 1971 is shown on Map R. This map may be compared with the similar map for 1964 (Map H). In addition to having a smaller number of farm households as a whole, in 1971 only 22.3 per cent of these made a specialty of farming. In 1971 it was only in Minami Daito, Aguni, and Tarama islands that farming was a primary activity for more than half of the farm households. On other isolated islands, and on Miyako, Yaeyama, and the Motobu Peninsula, significant percentages of the farm households made a specialty of farming, and even so this figure was still less than half the number of farm households. It is obvious that the urbanized areas of southern and central Okinawa contained few farm households which made a specialty of farming. As may be noted in Table 20, in 1971 only one-fourth of

28. A very interesting account of two cooperatives in northern Okinawa at Oku and Taira which were studied in the field is given in an article by William P. Lebra and Thomas W. Maretzki, "The Community Cooperative in Northern Okinawa," *Economic Development and Cultural Change* 11 (April 1963): 225–38.

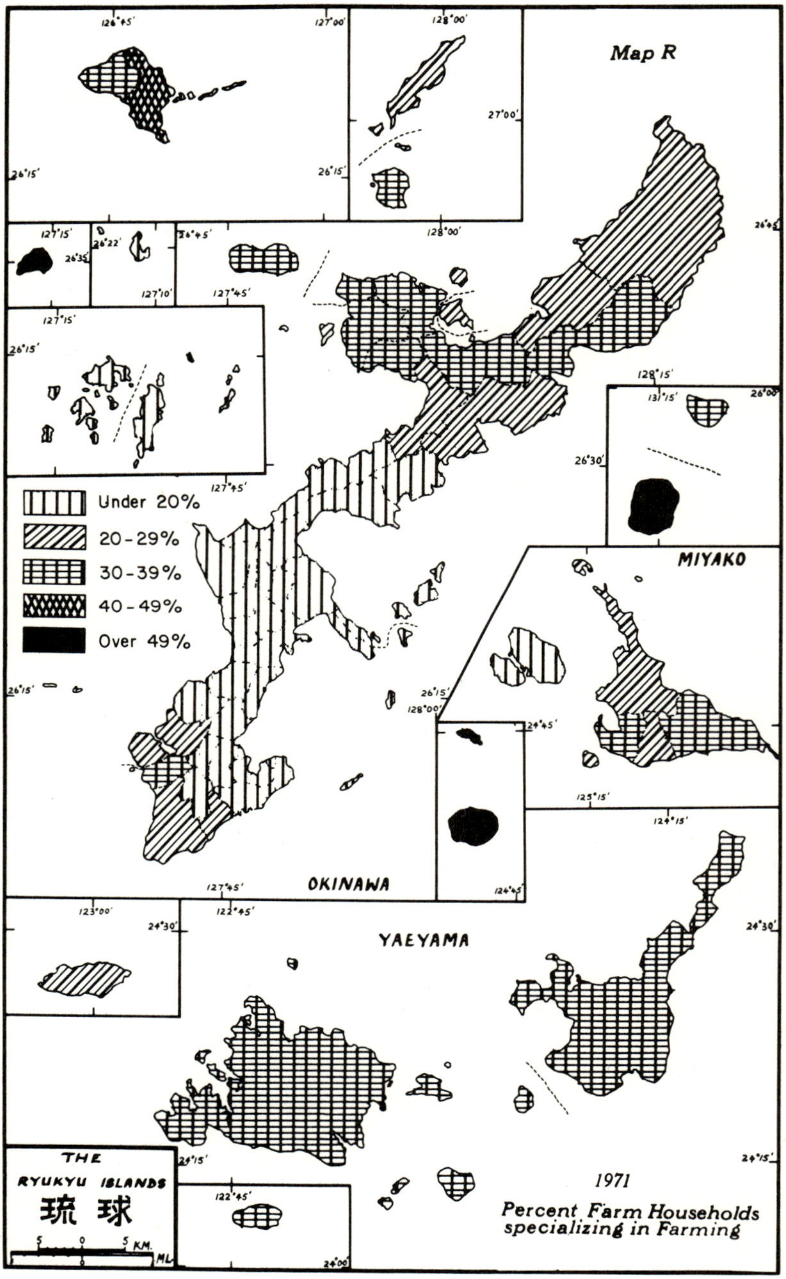

Map R
Under 20%
20-29%
30-39%
40-49%
Over 49%
OKINAWA
MIYAKO
YAEYAMA
THE
RYUKYU ISLANDS
琉球
KM.
ML.
1971
Percent Farm Households specializing in Farming

TABLE 20

ACTIVITIES OF MEN AND WOMEN OVER 15 YEARS OF AGE IN FARM HOUSEHOLDS, RYUKYU ISLANDS, 1971

Activity	Men	Women	Total	Per Cent
Engaged in farming full-time	24,174	28,243	52,417	25.2
Engaged in farming and in domestic duties		18,243	18,243	8.7
Engaged in farming, studying, and other activities	11,582	9,157	20,739	10.0
Engaged in farming primarily and in other employment	6,213	2,466	8,679	4.2
Engaged primarily in other employment with part-time farming	39,427	13,078	52,505	25.2
Engaged only in other employment	9,007	10,079	19,086	9.2
Not engaged in any employment, only in household activities		7,973	7,973	3.8
Not engaged in any employment, only in studying	6,411	9,680	16,091	7.7
Not engaged in any employment, only in other activities	3,974	8,482	12,456	6.0
Total	100,788	107,401	208,189	100.0

SOURCE: 1971 Agricultural Census, Government of the Ryukyu Islands.

the members of farm households over 15 years of age were engaged in full-time farming. Another quarter were engaged primarily in other employment while farming part time. Many of these were housewives and students.

Agriculture continued to be characterized by small scale farm households in the Ryukyu Islands in 1971. Since the decrease in farm households had been greater than the decrease in cultivated land between 1964 and 1971, there was some increase in cultivated land per farm household, but this was slight and the holdings per farm household were still quite small. The 58,683 operating farm households had under cultivation 116,463 acres of farm land, or an average of 2 acres per farm household. Table 21 gives data on the farm households and the amount of land they had under cultivation in 1964 and 1971. As may be noted there was a small shift toward larger holdings. It needs to be stressed, however, that most of the farms were small. In 1964, 43.1 per cent of the farm households, and in 1971, 40.1 per cent, had under cultivation between 10 and 50 *ares* (roughly one-fourth acre to one and one-fourth acres).

In addition to the farm households having land under cultivation, farming enterprises were carried on by companies and institutions. Some of these were schools which developed school

TABLE 21

SIZE OF LAND CULTIVATED BY FARM HOUSEHOLDS, RYUKYU ISLANDS, 1964, 1971 (in *ares*)

Size of Holdings	1964 Farm Households	1964 Per Cent	1971 Farm Households	1971 Per Cent
0–5	5,772	7.5	3,566	5.9
5–10	6,569	8.5	4,069	6.8
10–30	19,505	25.3	14,288	23.7
30–50	13,705	17.8	9,925	16.4
50–70	9,023	11.7	7,750	12.8
70–100	8,683	11.2	6,565	10.9
100–150	7,021	9.1	6,171	10.2
150–200	3,097	4.0	2,997	5.0
200–300	2,321	3.0	2,883	4.8
Over 300	1,433	1.9	2,132	3.5
Total	77,129	100.0	60,346	100.0

SOURCES: 1964 data derived from 1964 Agricultural Census, *1968 Okinawa Statistical Yearbook*, table on p. 110; 1971 data from 1971 Agricultural Census, Table 1.

gardens for educational purposes; some were government experimental stations. Though these enterprises had numbered 497 in 1964 there were only 265 such units in 1971. Yet the area which they cultivated decreased only slightly, from 2,885 acres to 2,734 acres. Fifty of these enterprises did not actually cultivate land; 57 cultivated more than 300 *ares* (7.4 acres) each. These large operations included 2 companies each of which had under cultivation more than 5,000 *ares* (123.5 acres). Of the total area of 2,734 acres, 46 per cent was in pineapple production and was owned by 28 companies; the largest of these holdings was in western Ishigaki.

11. Animal Husbandry in 1971

THE commercialization of the agriculture of the Ryukyu Islands, as was pointed out earlier, was accompanied by a decided decline in the farm animals kept on the farm for food and work. Yet at the same time a decided increase of commercial enterprises such as chicken and egg businesses, piggeries, and cattle "ranches" took place. Though some farms that have been developed for beef cattle are called in Japanese *ranchu,* they are in fact on a rather small scale. The available statistical materials from the *1971 Okinawa Statistical Yearbook* quantifies the considerable changes which have taken place. Table 14 gives data for 1971 on the livestock and poultry in the three major areas of the Ryukyu Islands and also includes data for 1939, 1952, and 1964.

The introduction of hand tractors and small trucks drastically reduced the need for horses in farm work. In 1971, there were only 6,137 horses as contrasted to the 14,396 horses in 1964. Over half of these were in Miyako. An even sharper decrease took place in the number of goats which had been grazed on the grass and forest lands in the rural areas, particularly of northern Okinawa and the offshore islands. More commercial and scientific farming resulted in a considerable increase of beef cattle, pigs, and chickens between 1964 and 1971. The Ryukyu Loan Development Corporation and other financing agencies gave loans for the development of relatively large scale operations. Dairy farms were established in Okinawa, particularly in the areas surrounding Naha, with a resultant increase in dairy cattle in Okinawa from 961 head in 1964 to 2,353 head in 1971. For dairy cattle as well as for pigs, beef cattle, and chickens, quality and production increased under more scientific management.

Maps S, T, U, and V show the distribution of dairy and beef cattle, swine, and chickens in 1971. Some new experiments were being conducted in extensive cattle "ranching" in northern Okinawa and in some of the offshore islands. Beef cattle had increased in number, particularly in Haneji-son and Kushi-son in northern Okinawa and on Ie Island. There was also an increase in beef cattle in Miyako and Yaeyama. In 1970, the Government of the Ryukyu Islands started a relatively large-scale beef cattle experiment and breeding station in northern Okinawa. In August of 1970 there

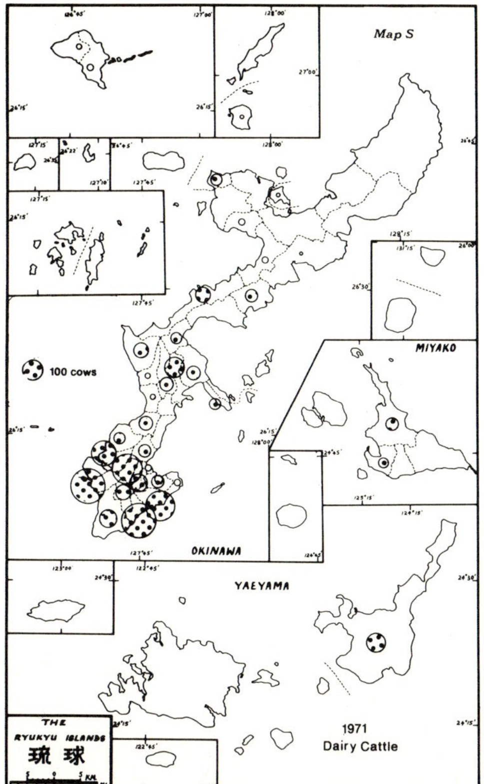
Map S
100 cows
MIYAKO
OKINAWA
YAEYAMA
THE
RYUKYU ISLANDS
琉球
1971
Dairy Cattle

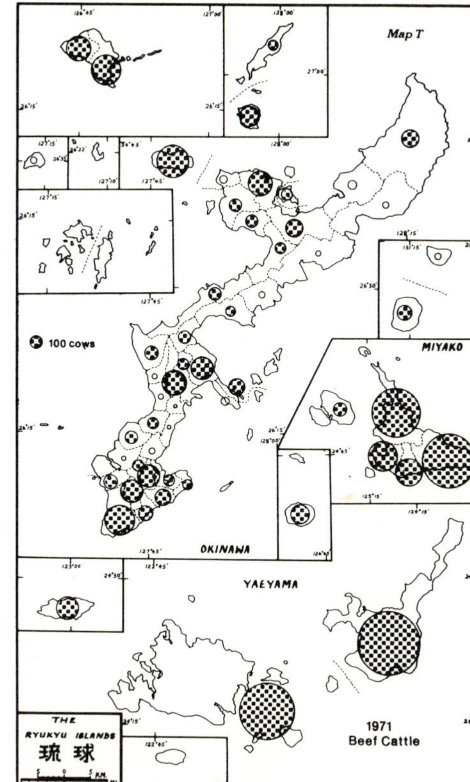
Map T
100 cows
MIYAKO
OKINAWA
YAEYAMA
THE
RYUKYU ISLANDS
琉球
1971
Beef Cattle

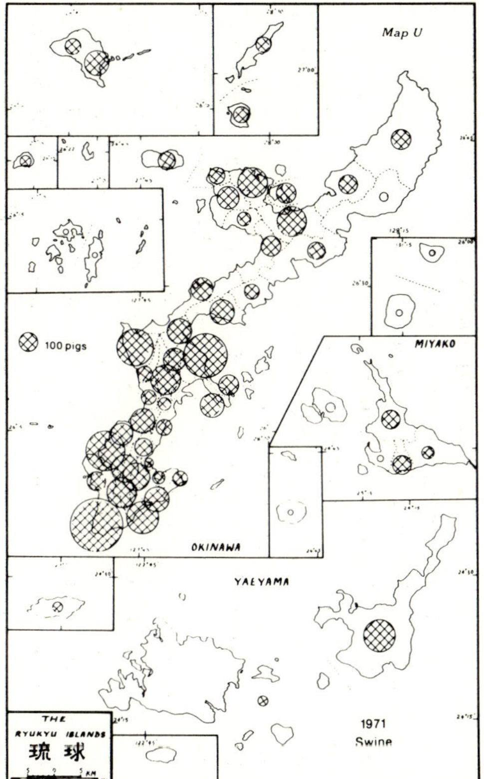
Map U
100 pigs
MIYAKO
OKINAWA
YAEYAMA
THE
RYUKYU ISLANDS
琉 球
1971
Swine

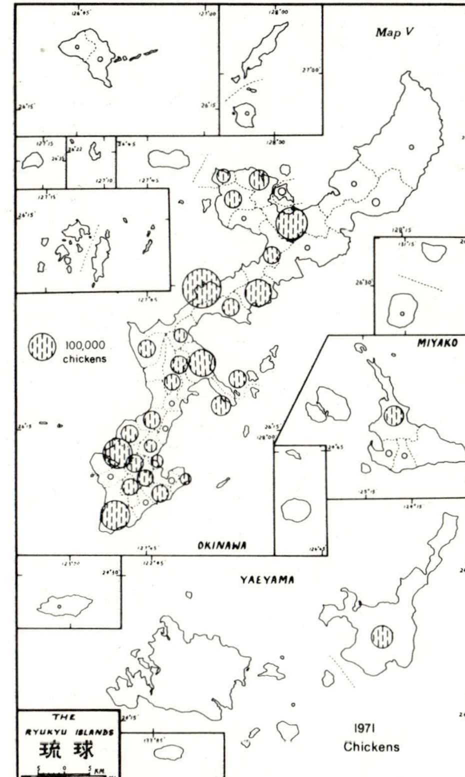
Map V
100,000
chickens
MIYAKO
OKINAWA
YAEYAMA
THE
RYUKYU ISLANDS
琉 球
1971
Chickens

were more staff members than heads of cattle, but the grazing lands had been cleared, water supplies piped in, and dipping tanks installed. It was hoped that this type of relatively extensive land use for beef cattle production, if successful, might be considerably expanded in the hilly areas of northern Okinawa, the Motobu Peninsula, and Yaeyama. Some hoped that the extensive area held by the United States Armed Forces for military training grounds might be released for cattle "ranches."

The raising of pigs has had a long tradition in the Ryukyu Islands. Meals with pork are prized for special occasions or served in restaurants. Thus, in contrast with some areas of Asia, the expansion of swine production is based both on agricultural practices and on increased markets in a more affluent society. The numbers of swine for the Ryukyu Islands as a whole increased from 145,614 in 1964 to 187,128 in 1971. The numbers were even higher in 1969 and 1970. There has been a noticeable trend away from the barnyard pigpen to modern piggeries with cement block walls, adequately ventilated, and kept moderately clean. These modern piggeries are now found in central Okinawa, in the southern part of northern Okinawa, and on the Motobu Peninsula. The expansion has been significant in some of these areas. Development of modern swine production continued near the city of Naha, reflecting the increased market for pork products. In contrast to these developments, swine production decreased in Miyako and increased only slightly in Yaeyama. Throughout the Ryukyu Islands, because many small producers still fed their swine sweet potatoes, some correlation continued between areas of sweet potato production and swine production, but increasingly production was being concentrated in modern operations.

The same pattern of commercialization and concentration of production in larger units has taken place in the chicken and egg production of the Ryukyu Islands. Haneji-son was an area in which chickens were produced commercially in relatively large establishments. In 1970, epidemics of Newcastle disease and Marek's disease necessitated the destruction of some of the flocks in this area of northern Okinawa. The influence of the market on production resulted in an increase in commercial production of chickens and eggs, including some modern hatcheries and brooder houses, near Naha and the Kadena Air Base. There are expectations that this type of commercial operation will continue to expand.

12. Future

THE agriculture of the Ryukyu Islands has gone through many significant changes. It has moved from a dominant place in the economy to one of secondary importance. The attitudes of many people toward farming tend to be nostalgic rather than realistic. Farming is considered by them as a proper Okinawan way of life. Yet part-time farming has increased greatly while the number of persons who made a specialty of farming has decreased drastically. The commercialization of agriculture and the export of sugar and pineapples, greatly dependent upon high Japanese subsidies, have become paramount. Similarly, vegetable production, cattle raising, and commercial production of swine and chickens are dependent upon selling these products to urban dwellers, to the United States Armed Forces, and to those establishments that cater to tourists. The self-sufficient "family farm" is passing rapidly from the Okinawan scene.

There has been an increasing geographical concentration of certain crops such as sugar and pineapples on certain soils. Sugar is concentrated on the limestone-based soils of southern Okinawa, Miyako, and the Daito Islands; pineapple on the acid, deep soils in northern Okinawa and Ishigaki. Dairying, intensive vegetable growing, and relatively large-scale piggeries are located near the urban centers of the Naha-Koza area. In many areas, such as those on the offshore islands, once-productive agricultural lands have been abandoned because of the large-scale out-migration of their young population to seek a better and more attractive livelihood.

With the reversion of Okinawa to Japan, with the probable decrease of the United States Armed Forces, and with the great increase in tourism, profound changes will undoubtedly come to Okinawa Prefecture and to the agricultural way of life of its people. It is difficult to predict these changes. Certainly, new patterns of land ownership, enlargement of the scale of agricultural production, and subsidized marketing are emerging. These and other developments point to continued profound modification in Okinawan agriculture.

Appendixes

A. Research and Information Papers, Ryukyu Islands Project

AS a corollary of a major research project on the Ryukyu Islands, 300–400 copies of a series of papers have been prepared in xeroxed form. Some of these include the reproduction of old documents which have been edited with introductions and notes. Some are reports on research and some are the preliminary results of research. Copies of these *Research and Information Papers* have been distributed to various research libraries and to scholars interested in the Ryukyu Islands. They are listed below:

1. A Proposed Research Project on Agricultural Changes in the Ryukyu Islands: Geographical Aspects (14 pages, May 1, 1970).
2. A Bibliography of Useful English Language Materials on the Ryukyu Islands (16 pages, Nov. 1, 1970).
3. The Ryukyus: Islands in a Strategic Location (17 pages, Dec. 1, 1970).
4. The Ryukyus: Islands in Transition (24 pages, Dec. 15, 1970).
5. Progress Report on a Research Project on Agricultural Changes in the Ryukyu Islands: Geographic Aspects (19 pages, Jan. 1, 1971).
6. Basil Hall Chamberlain's Writings on the Ryukyu Islands (15 pages, May 1, 1971).
7. The Luchu Islands and Their Inhabitants, *The Geographical Journal,* 1895 by Basil Hall Chamberlain (33 pages, June 1, 1971).
8. Notes on Basil Hall Chamberlain's "The Luchu Islands and Their Inhabitants" (23 pages, May 1, 1971).
9. Migration from Hamahiga Island by Tomonari Ishikawa (26 pages, Jan. 10, 1972).
10. The Islands of the Ryukyus (30 pages, Jan. 20, 1972).

11. The Senkaku Islands (21 pages, Jan. 30, 1972).
12. George Jones' Account of the Geology of Okinawa, 1853–1854 (37 pages, Feb. 10, 1972).
13. D. S. Green's Report on the Agriculture of Okinawa, 1853–1854 (18 pages, Feb. 10, 1972).
14. James Morrow's Observations on the Agriculture of Okinawa, 1854 (31 pages, Feb. 10, 1972).
15. C. F. Fahs' Account of Okinawa, 1854 (14 pages, Feb. 10, 1972).
16. Geographical Papers on the Ryukyu Islands (29 pages, Nov. 1972).
17. Fishing Activities in the Ryukyu Islands (28 pages, Dec. 1972).
18. Sangoku Tsuran Zusetsu: A 1785 Account of the Ryukyu Islands (60 pages, Nov. 1972).
19. Civilian Casualties in the Battle of Okinawa by Norman D. King (11 pages, Nov. 1972).
20. The Geologic Base of the Ryukyu Islands (19 pages, April 1973).
21. Physiographic Regions of the Ryukyu Islands: Okinawa (35 pages, April 1973).
22. Physiographic Regions of the Ryukyu Islands: Sakishima (26 pages, April 1973).
23. The Climate of the Ryukyu Islands (37 pages, April 1973).

B. Computer-drawn Maps of the Ryukyu Islands

IN order to assist this study of the geographical aspects of agricultural changes in the Ryukyu Islands, a computer-drawn map of the islands was developed by Steven T. Gladin, a graduate student in the Department of Geography of the University of Florida. The map occupies a double computer sheet, 12.5 by 21 inches. A copy of the map is reproduced on a much reduced scale and a portion of the map that includes southern Okinawa is reproduced at close to the scale of the original.

The map is made up of 12 inset maps formed into three large groups: Okinawa, Miyako, and Yaeyama. Each of the 12 inset maps

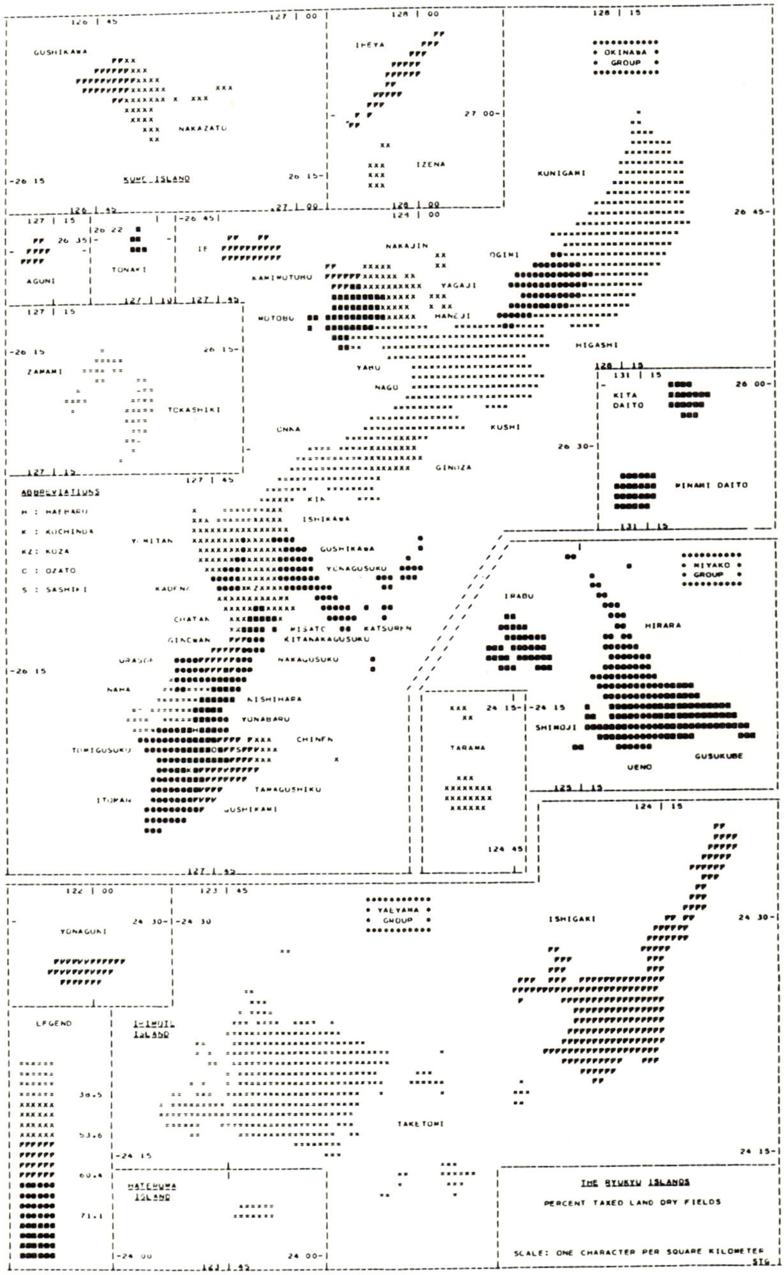

OKINAWA GROUP
KUME ISLAND
GUSHIKAWA
NAKAZATO
IHEYA
IZENA
KUNIGAMI
AGUNI
TONAKI
IE
NAKAJIN
OGIMI
KAMIMOTOBU
YAGAJI
MOTOBU
HANEJI
HIGASHI
ZAMAMI
TOKASHIKI
YABU
NAGO
KUSHI
ONNA
GINOZA
KITA DAITO
MINAMI DAITO
ABBREVIATIONS
H : HAEBARU
K : KOCHINDA
KZ: KOZA
O : OZATO
S : SASHIKI
KIN
ISHIKAWA
YOMITAN
GUSHIKAWA
YONAGUSUKU
KADENA
CHATAN
MISATO
KATSUREN
GINOWAN
KITANAKAGUSUKU
URASOE
NAKAGUSUKU
NAHA
NISHIHARA
YONABARU
CHINEN
TOMIGUSUKU
TAMAGUSUKU
ITOMAN
GUSHIKAMI
MIYAKO GROUP
IRABU
HIRARA
TARAMA
SHIMOJI
UENO
GUSUKUBE
YAEYAMA GROUP
YONAGUNI
ISHIGAKI
LEGEND
36.5
53.6
60.4
71.1
IRIOMOTE ISLAND
TAKETOMI
HATERUMA ISLAND
THE RYUKYU ISLANDS
PERCENT TAXED LAND DRY FIELDS
SCALE: ONE CHARACTER PER SQUARE KILOMETER

ABBREVIATIONS
H : HAEBARU
K : KOCHINDA
KZ: KOZA
O : OZATO
S : SASHIKI
127 | 45
-26 15
GINOZA
KIN
ISHIKAWA
YOMITAN
GUSHIKAWA
YONAGUSUKU
KADENA
CHATAN
MISATO
KATSUREN
GINOWAN
KITANAKAGUSUKU
URASOE
NAKAGUSUKU
NAHA
NISHIHARA
YONABARU
CHINEN
TOMIGUSUKU
TAMAGUSHIKU
ITOMAN
GUSHIKAMI
LEGEND
38.5
53.6
60.4
71.1
THE RYUKYU ISLANDS
PERCENT TAXED LAND DRY FIELDS
SCALE: ONE CHARACTER PER SQUARE KILOMETER
SIG

has the coordinates of latitude and longitude printed on it. The symbols are printed in a close approximation of the shape of the 43 inhabited islands of the Ryukyus; the uninhabited islands are not included. The 59 political and statistical units into which the Ryukyu Islands were divided in 1970 are named and programmed so that appropriate data may be printed in their area. In some cases, for example, Taketomi-son in the Yaeyama group, a number of islands of varying size are included in the one political area or *son*. In the case of Okinawa, there are a number of small political units: cities, towns, and *son*; the five interior areas of southern Okinawa have one or two initials abbreviating the name of the political unit placed within the area and the abbreviations and the full names of the political units are printed on the left margin of the map. The symbols in which the data are presented are each equivalent to one square kilometer in area, thus providing an automatic areal scale to the map. This is a unique feature not found on most computer-drawn maps.

The computer program was developed so that data from each of the political and statistical units could be card-punched and then run through the computer for the needed calculations, for arrangements in categories, and for printing the results. The data may be arranged in quintiles, in standard deviations, or in any pre-determined five categories. The example given of the southern Okinawa segment of the map shows the percentage in 1964 of the taxed land cultivated by farm households which was devoted to dry fields. Computer-drawn maps may be used for any analysis of data ranging from percentage changes in specific crops to relationships between crops; population data of various kinds may be easily computed and drawn on the map.

Once the data have been put on cards, the computer-drawn map may be computed and printed relatively cheaply. One group of 40 maps, for example, were done in 12.36 seconds, so that costs for computer time were less than 20 cents per map. A computer-drawn map such as this has many advantages for research on agricultural and other economic and demographic changes in areas such as the Ryukyu Islands. It enables one to compute and visualize geographical shifts in production and in distributions.

C. American-Sponsored Research Studies of the Ryukyu Islands in the 1950s

THERE were a number of research studies on the Ryukyu Islands made in the 1950s. These were supported by the direct or indirect financial assistance of the United States Army which had responsibility for the Civil Administration of the Ryukyu Islands. In 1950–51, the Amami-Oshima Islands, the northern part of the Ryukyu Island arc, were under the Civil Administration of the United States. On December 25, 1953, their control was transferred to Japan and they again became a part of Kagoshima Prefecture under whose political control they had been after 1609. Thus, the excellent studies by Douglas Haring on the Amami-Oshima are not included in this bibliographical note, though they may be read with profit for comparative purposes.

Among the basic research studies were those on the military geology of the Ryukyu Islands. The results of these studies were published in subsequent years. The titles of these significant studies are included in an appendix: Bibliography of English Language Materials on the Geology of the Ryukyu Islands, in *Research and Information Paper*, Ryukyu Islands Project, number 20 (April 1973).

With financial assistance from the United States Army, the Pacific Science Board of the National Academy of Sciences prepared a number of reports under a project entitled Scientific Investigations in the Ryukyu Islands (SIRI). These included geographical and anthropological studies of various villages and have special value in terms of an analysis of the geographical aspects of agriculture in the Ryukyu Islands in this period. Clarence J. Glacken, a geographer, studied three villages in 1951: Hanashiro and Minatogawa in southern Okinawa, and Matsuda in northern Okinawa. (These are located on Map W that accompanies this appendix.) His SIRI report on these villages was the basis for a book he subsequently published, which also includes some insightful general information on the Ryukyu Islands as of this period: *The Great Loochoo, A Study of Okinawan Village Life* (Berkeley and Los Angeles: University of California Press, 1955). A paper-back edition of this was published by Charles E. Tuttle Company, Rutland, Vermont, and Tokyo, Japan, 1960.

F. R. Pitts, W. P. Lebra, and W. P. Suttles, a geographer and

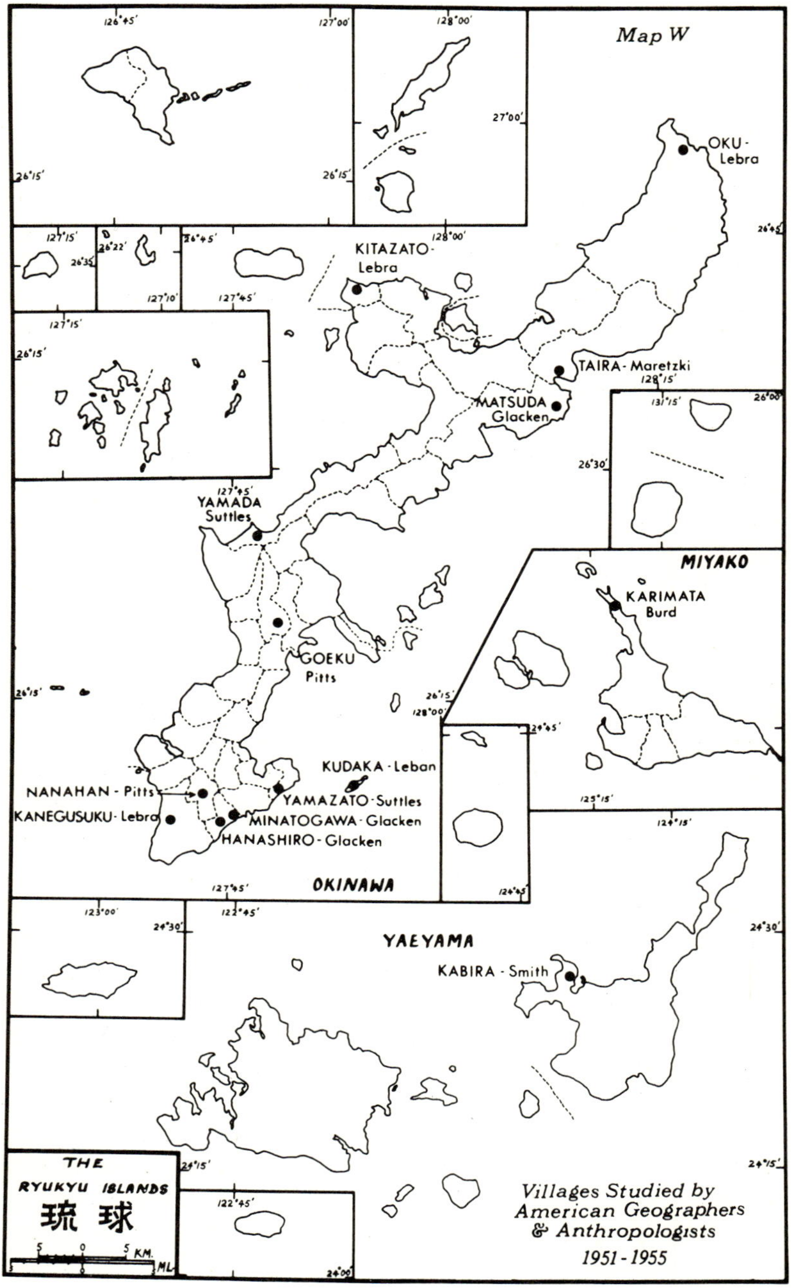

Map W
OKU - Lebra
KITAZATO - Lebra
TAIRA - Maretzki
MATSUDA Glacken
YAMADA Suttles
GOEKU Pitts
KUDAKA - Leban
NANAHAN - Pitts
YAMAZATO - Suttles
KANEGUSUKU - Lebra
MINATOGAWA - Glacken
HANASHIRO - Glacken
OKINAWA
MIYAKO
KARIMATA Burd
YAEYAMA
KABIRA - Smith
THE RYUKYU ISLANDS
琉球
KM.
ML.
Villages Studied by American Geographers & Anthropologists 1951 - 1955

two anthropologists, wrote a general report, "Post-War Okinawa," mimeographed SIRI Report no. 8 (Washington, D.C.: Pacific Science Board, National Research Council, June 1955). This was based in part on village studies by Pitts (Nanahan, or Tomoyose, and Goeku), Lebra (Kanegusuku and Kitazato), and Suttles (Yamazato and Yamada) which were carried out in 1953 and 1954.

William W. Burd did a study of Karimata in Miyako in 1951. The results were published in SIRI Report no. 3, mimeographed (Washington, D.C.: Pacific Science Board, National Research Council, September 1951), and are available in more finished form in "Karimata, A Village in the Southern Ryukyu," typescript (Berkeley, 1952).

Dr. and Mrs. Allan H. Smith did a study of Kabira in Ishigaki. Some of the results are in SIRI Report no. 1, mimeographed (Washington, D.C.: Pacific Science Board, National Research Council), and a brief summary was published in Allan H. Smith, "The Culture of Kabira, Southern Ryukyu Islands," *Proceedings, American Philosophical Society* 104 (April 1960): 134–71.

Subsequent to the studies made as a part of the SIRI project two other village studies were made. One was on the land tenure system on Kudaka Island by Richard W. Lieban; the result was a Ph.D. dissertation submitted in 1956 to Columbia University. Another study was done in 1954 by Thomas W. Maretzki and Hatsumi Maretzki of Taira in northern Okinawa; this was published as *Taira: An Okinawan Village,* Six Cultures Series, vol. 7 (New York: John Wiley and Sons, Inc., 1966).

Another study sponsored by the United States Army was carried on without any extended field research by M. H. Nuttonson of the American Institute of Crop Ecology which published it. This study includes climatic, geologic, and natural vegetation data collected from various sources. It concludes with a section on Farm Crops and Farm Practices relying mainly on United States Civil Administration documents and other reports. Comparisons are made with areas of similar ecological conditions in the Northern Hemisphere, mainly Louisiana, Alabama, and Florida. It was published in 1952 in Washington, D.C. under the authorship of Nuttonson and entitled *Ecological Crop Geography and Field Practices of the Ryukyu Islands, Natural Vegetation of the Ryukyus, and Agro-Climatic Analogues in the Northern Hemisphere.*

Dr. Egbert H. Walker of the Smithsonian Institution, working with three Ryukyuan botanists, studied the flora of the Ryukyu Islands. A limited number of an enumeration, *Flora of Okinawa,* were reproduced by the United States Civil Administration of the Ryukyu Islands in cooperation with the Forestry Bureau, Department of Natural Resources, Government of the Ryukyu Islands, in 1952. Walker also prepared a report, *Important Trees of the Ryukyu Islands,* but copies of this are very scarce. From this was derived a publication, *Ryukyu Islands: Preliminary Notes on the Use, Distribution and Adaptability of Native and Introduced Tree Species,* Special Bulletin no. 1 (Naha: United States Civil Administration of the Ryukyu Islands, November 1952). His major work was an exhaustive study of many years of dedicated effort of the flora of the Ryukyu Islands. This is in the files of the Smithsonian Institution in Washington and has not yet been published.

UNIVERSITY OF FLORIDA MONOGRAPHS

Social Sciences

1. *The Whigs of Florida, 1845–1854,* by Herbert J. Doherty, Jr.

2. *Austrian Catholics and the Social Question, 1918–1933,* by Alfred Diamant

3. *The Siege of St. Augustine in 1702,* by Charles W. Arnade

4. *New Light on Early and Medieval Japanese Historiography,* by John A. Harrison

5. *The Swiss Press and Foreign Affairs in World War II,* by Frederick H. Hartmann

6. *The American Militia: Decade of Decision, 1789–1800,* by John K. Mahon

7. *The Foundation of Jacques Maritain's Political Philosophy,* by Hwa Yol Jung

8. *Latin American Population Studies,* by T. Lynn Smith

9. *Jacksonian Democracy on the Florida Frontier,* by Arthur W. Thompson

10. *Holman Versus Hughes: Extension of Australian Commonwealth Powers,* by Conrad Joyner

11. *Welfare Economics and Subsidy Programs,* by Milton Z. Kafoglis

12. *Tribune of the Slavophiles: Konstantin Aksakov,* by Edward Chmielewski

13. *City Managers in Politics: An Analysis of Manager Tenure and Termination,* by Gladys M. Kammerer, Charles D. Farris, John M. DeGrove, and Alfred B. Clubok

14. *Recent Southern Economic Development as Revealed by the Changing Structure of Employment,* by Edgar S. Dunn, Jr.

15. *Sea Power and Chilean Independence,* by Donald E. Worcester

16. *The Sherman Antitrust Act and Foreign Trade,* by Andre Simmons

17. *The Origins of Hamilton's Fiscal Policies,* by Donald F. Swanson

18. *Criminal Asylum in Anglo-Saxon Law,* by Charles H. Riggs, Jr.

19. *Colonia Barón Hirsch, A Jewish Agricultural Colony in Argentina,* by Morton D. Winsberg

20. *Time Deposits in Present-Day Commercial Banking,* by Lawrence L. Crum

21. *The Eastern Greenland Case in Historical Perspective,* by Oscar Svarlien

22. *Jacksonian Democracy and the Historians,* by Alfred A. Cave

23. *The Rise of the American Chemistry Profession, 1850–1900,* by Edward H. Beardsley

24. *Aymara Communities and the Bolivian Agrarian Reform,* by William E. Carter

25. *Conservatives in the Progressive Era: The Taft Republicans of 1912,* by Norman M. Wilensky

26. *The Anglo-Norwegian Fisheries Case of 1951 and the Changing Law of the Territorial Sea,* by Teruo Kobayashi

27. *The Liquidity Structure of Firms and Monetary Economics,* by William J. Frazer, Jr.

28. *Russo-Persian Commercial Relations, 1828–1914,* by Marvin L. Entner

29. *The Imperial Policy of Sir Robert Borden,* by Harold A. Wilson

30. *The Association of Income and Educational Achievement,* by Roy L. Lassiter, Jr.

31. *Relation of the People to the Land in Southern Iraq,* by Fuad Baali

32. *The Price Theory of Value in Public Finance,* by Donald R. Escarraz

33. *The Process of Rural Development in Latin America,* by T. Lynn Smith

34. *To Be or Not to Be . . . Existential-Psychological Perspectives on the Self,* edited by Sidney M. Jourard

35. *Politics in a Mexican Community,* by Lawrence S. Graham

36. *A Two-Sector Model of Economic Growth with Technological Progress,* by Frederick Owen Goddard

37. *Florida Studies in the Helping Professions,* by Arthur W. Combs

38. *The Ancient Synagogues of the Iberian Peninsula,* by Don A. Halperin

39. *An Estimate of Personal Wealth in Oklahoma in 1960,* by Richard Edward French

40. *Congressional Oversight of Executive Agencies,* by Thomas A. Henderson

41. *Historians and Meiji Statesmen,* by Richard T. Chang

42. *Welfare Economics and Peak Load Pricing: A Theoretical Application to Municipal Water Utility Practices,* by Robert Lee Greene

43. *Factor Analysis in International Relations: Interpretation, Problem Areas, and an Application,* by Jack E. Vincent

44. *The Sorcerer's Apprentice: The French Scientist's Image of German Science, 1840–1919,* by Harry W. Paul

45. *Community Power Structure: Propositional Inventory, Tests, and Theory,* by Claire W. Gilbert

46. *Human Capital, Technology, and the Role of the United States in International Trade,* by John F. Morrall III

47. *The Segregation Factor in the Florida Democratic Gubernatorial Primary of 1956,* by Helen L. Jacobstein

48. *The Navy Department in the War of 1812,* by Edward K. Eckert

49. *Social Change and the Electoral Process,* by William L. Shade

50. *East from the Andes: Pioneer Settlements in the South American Heartland,* by Raymond E. Crist and Charles M. Nissly

51. *A General Equilibrium Study of the Monetary Mechanism,* by David L. Schulze

52. *The Impact of Trade Destruction on National Incomes: A Study of Europe 1924–1938,* by Philip Friedman

53. *Administration of Justice in Drunk Driving Cases,* by Joseph W. Little

54. *Geographical Aspects of Agricultural Changes in the Ryukyu Islands,* by Shannon McCune